Proceedings
of the
Steklov Institute of Mathematics

1978, Issue 4

C-types of n-dimensional lattices and 5-dimensional primitive parallelohedra (with application to the theory of coverings)

by

S. S. Ryškov and E. P. Baranovskiĭ

Translation of

ТРУДЫ
ордена Ленина
МАТЕМАТИЧЕСКОГО ИНСТИТУТА
имени В. А. СТЕКЛОВА

137 (1976)

АКАДЕМИЯ НАУК

СОЮЗА СОВЕТСКИХ СОЦИАЛИСТИЧЕСКИХ РЕСПУБЛИК

ТРУДЫ

ордена Ленина

МАТЕМАТИЧЕСКОГО ИНСТИТУТА

имени В. А. СТЕКЛОВА

CXXXVII

C-ТИПЫ *n*-МЕРНЫХ РЕШЕТОК И ПЯТИМЕРНЫЕ ПРИМИТИВНЫЕ ПАРАЛЛЕЛОЭДРЫ (С ПРИЛОЖЕНИЕМ К ТЕОРИИ ПОКРЫТИЙ)

С. С. РЫШКОВ и Е. П. БАРАНОВСКИЙ

Ответственный редактор академик С. М. НИКОЛЬСКИЙ

Издательство "Наука"
Москва 1976

Translated by R. M. Erdahl

AMS (MOS) subject classifications (1970). Primary 10E30; Secondary 10E25.

ABSTRACT. This monograph is devoted to one of the domains of discrete geometry. A new approach is given to the classification problem for n-dimensional parallelohedra as well as a solution of this problem for $n = 5$. As an application the problem of least dense lattice covering of five-dimensional Euclidean space by equal spheres is solved.

Library of Congress Cataloging in Publication Data

Ryshkov, Sergeĭ Sergeevich.

C-types of n-dimensional lattices and 5-dimensional primitive parallelohedra.

(Proceedings of the Steklov Institute of Mathematics; no. 137 (1976))

Translation of C-tipy n-mernykh reshetok i piatimernye primitivnye paralleloedry.

Bibliography: p.

1. Geometry of numbers. 2. Lattice theory. 3. Form, Quadratic. I. Baranovskiĭ, Evgeniĭ Petrovich, 1924– joint author. II. Title. III. Series: Akademii͡a nauk SSSR. Matematicheskiĭ institut. Proceedings; no. 137.

QA1.A413 no. 137 [QA241.5] 510′.8s [512′.72]

ISBN 0-8218-3037-6 78-21923

Translation authorized by the
All-Union Agency for Author's Rights, Moscow

PROCEEDINGS OF THE STEKLOV INSTITUTE OF MATHEMATICS IN THE ACADEMY OF SCIENCES OF THE USSR

TABLE OF CONTENTS

RUSSIAN TABLE OF CONTENTS

INTRODUCTION

The present work contains some geometric results stated in the language of quadratic forms. Both the main body of text and the supplementary chapter (Chapter IV) are prefaced by geometrical introductions containing some information of an historical character.

The main results of this work are as follows:

(1) The concept of the C-type of a lattice is introduced, and an algorithm for investigating the C-types of all n-dimensional lattices is constructed. Based on this we could construct a new algorithm for deducing the L-types (Voronoĭ types) of all n-dimensional lattices.

(2) All the general C-types and L-types of 5-dimensional lattices are found, as well as all of the 5-dimensional primitive parallelohedra.

(3) The relationship between the so-called simple subdivisions of the positive cone is found. In particular, the connection between the first and second subdivision of Voronoĭ is discovered [1], [2], giving thereby answers to some of the questions proposed by Voronoĭ in those works. In the chapter on applications the problems of the most economical lattice covering the 5-dimensional space by equal spheres is solved.

This work gives a detailed presentation of the results announced by the authors in [3]–[6].

1. Definitions. History of the questions. For the study of lattices of crystalline substances the famous Russian crystallographer E. S. Fedorov [7] found, by purely geometrical methods, two types of two-dimensional (Figure 1) and five types of three-dimensional (Figure 2) parallelohedra which, with the help of parallel translations, are capable of filling the entire plane (space) without gaps or overlapping. He showed that there are no other such parallelohedra for Euclidean spaces of dimension 2 or 3.

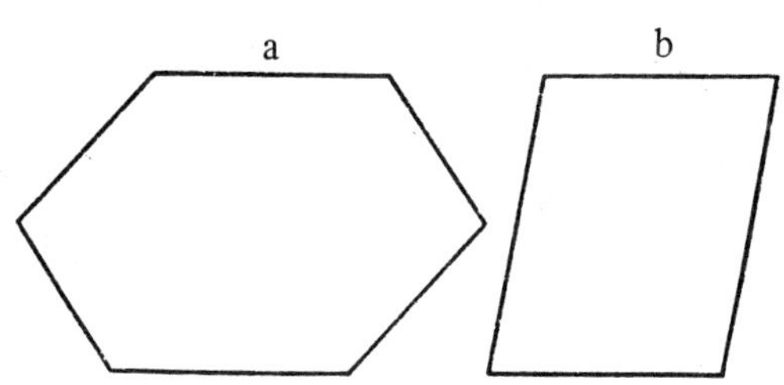

FIGURE 1

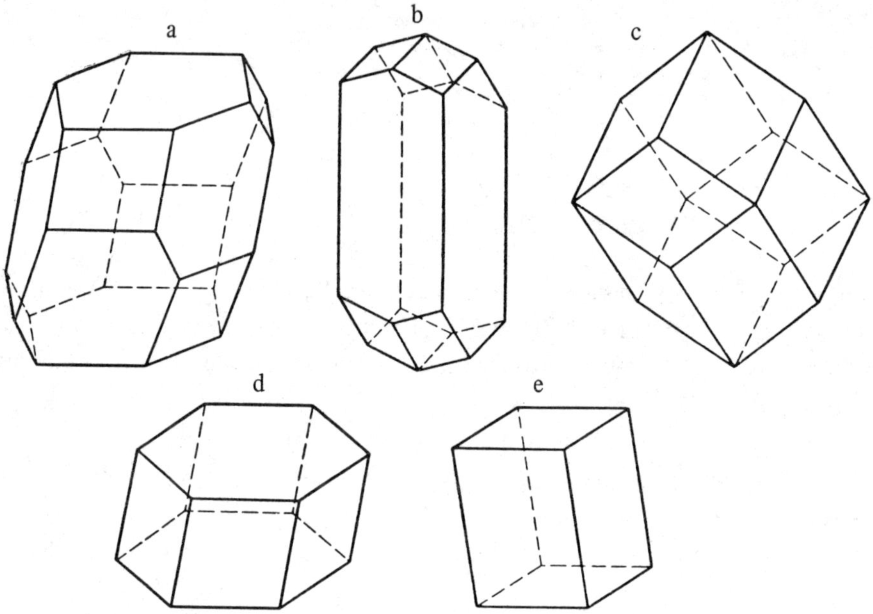

FIGURE 2

A covering of space by means of a given parallelohedron has been called a "filling of space" by Fedorov and the modern crystallographers following him. However, we will adopt the expression "a subdivision of space into parallelohedra", meaning in addition "by congruent and parallel arrangements", a qualifying phrase which we will drop. By a "subdivision of space" we will mean a covering in which the covering sets are closed and have pairwise disjoint interiors.

The subdivision of n-dimensional space into convex polyhedra, in particular into parallelohedra, may be such that the intersection of any two polyhedra is either empty or composed of an entire face, the dimension of which is arbitrary. In this case the subdivision and the parallelohedron defining the subdivision are called normal. If the subdivision (into parallelohedra) has contiguities which do not consist of entire faces, then the subdivision (and the parallelohedron) is called nonnormal.

For example, the two-dimensional parallelohedron of Figure 1a yields only normal subdivisions, i.e., it is always normal. The parallelohedron of Figure 1b may appear either as a normal (Figure 3) or as a nonnormal (Figure 4) parallelohedron. In the three-dimensional case the parallelohedra of Figures 2a, b, c are always normal. The parallelohedra of Figures 2d, e may be either normal or nonnormal.

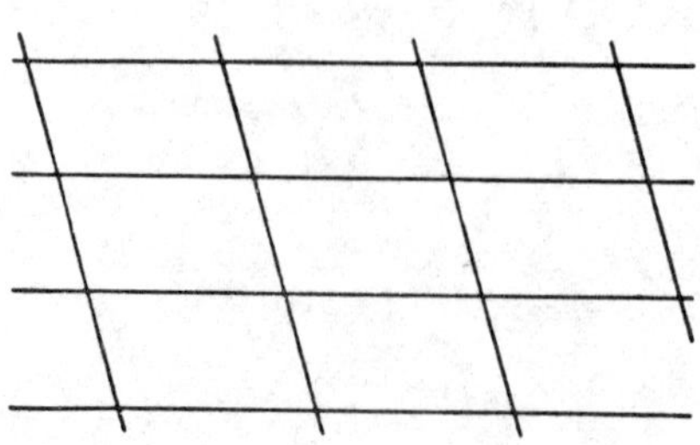

FIGURE 3

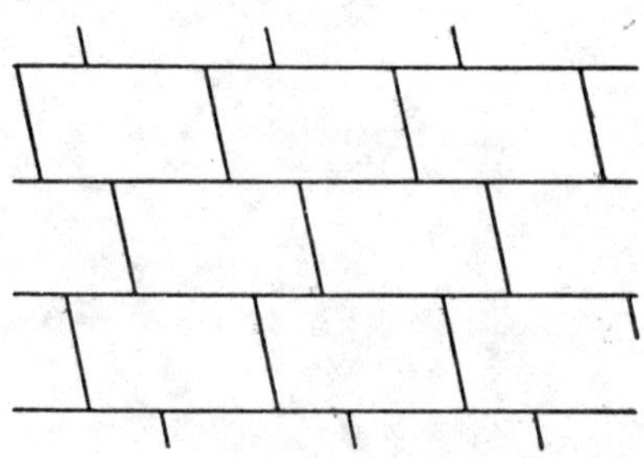

FIGURE 4

For larger dimensions we only know [8] that 4-dimensional nonnormal parallelohedra may also appear as normal. Throughout this monograph we will examine only normal parallelohedra.

Among the parallelohedra of Figures 1 and 2 only the hexagon (1a) and the 14-faced (2a) have the property that in their corresponding subdivisions the minimum possible number of parallelohedra meet at a vertex of the subdivision; in the plane three parallelohedra meet at a vertex and in space four parallelohedra meet at each vertex. Parallelohedra corresponding to subdivisions having this property are called primitive. The subdivision themselves are called primitive.

We will consider the most important class of n-dimensional parallelohedra. Let n-dimensional Euclidean space E^n be given the frame $E = (\bar{e}_1, \ldots, \bar{e}_n)$ and let the lattice Γ be the union of the integer points with respect to this frame. The frame E, as well as all other frames yielding the same lattice Γ, are called fundamental frames for the lattice. The domain of action of a point $A \in \Gamma$ with respect to the lattice Γ consists of the set of all points of E^n whose distance from A is not greater than that from any other point of the lattice.

The domain of action (but only for two-dimensional lattices) was introduced by Dirichlet. Voronoĭ's famous memoir [2] is devoted to a systematic investigation of these domains of action in the n-dimensional case. These domains of action have been called Voronoĭ domains and Dirichlet domains. Here we will call them Voronoĭ-Dirichlet domains and the corresponding subdivisions Voronoĭ-Dirichlet subdivisions.

In [2], first it was shown that the Voronoĭ-Dirichlet domains of points of an n-dimensional lattice are bounded, convex, finite-sided polyhedra and therefore parallelohedra (see also [10]). Second, the basic theorem in the theory of parallelohedra was stated: each n-dimensional primitive parallelohedron is the affine image of a Voronoĭ-Dirichlet domain for some n-dimensional lattice defined by the given parallelohedron. (Subsequently it was shown [11], it least for $n \leqslant 4$, that this property is possessed by not only the primitive parallelohedra but by all parallelohedra. For a further strengthening of this basic theorem see [12].)

The following important notions in the theory of parallelohedra were introduced by Voronoĭ [2]. He introduced the L-subdivision, the L-polyhedron and the L-type (or Voronoĭ type, or simply type) of an n-dimensional parallelohedron which is the affine image of the Voronoĭ-Dirichlet domain of some lattice. We cite a definition of L-subdivision, based on an idea contained in work of B. N. Delone [13], [14]. Namely a convex polyhedron with vertices at the points of the lattice is called an L-polyhedron of the lattice if it possesses the following two properties:

1. There exists a sphere circumscribed around the polyhedron.
2. The circumscribed sphere (closed) contains only those lattice points which occur as vertices of the polyhedron.

It has been shown (see for example [10]) that the collection of L-polyhedra of a given lattice forms a normal subdivision of the space of the lattice. The L-subdivision of a lattice is defined to be this normal subdivision of the space E^n of the lattice. The L-subdivision is dual in the usual combinatorial-metrical sense to the Voronoĭ-Dirichlet subdivison. Strictly speaking, this fact served as the basis of the first definition of L-subdivison [2].

Note that if the Voronoĭ-Dirichlet subdivision is primitive, then the corresponding L-subdivision is simplicial. Lattices having primitive Voronoĭ-Dirichlet subdivisions are called primitive; the others, special. The Voronoĭ-Dirichlet subdivisions corresponding to two distinct n-dimensional lattices, and also the lattices themselves, will be considered to belong to the same geometrical L-type if the corresponding L-subdivisions are affinely equivalent. This definition carries over to parallelohedra which are the affine images of Voronoĭ-Dirichlet domains of lattices and thereby, in view of the basic theorem in the theory of parallelohedra, to arbitrary primitive parallelohedra. When a parallelohedron belongs to an L-type, this determines the combinatorial structure of the parallelohedron as a polyhedron.

Voronoĭ [2] showed that for each value of n there are only a finite number of L-types which correspond to primitive n-dimensional parallelohedra, and constructed an algorithm to deduce these types. Using his algorithm he found and in great detail described all the types of primitive parallelohedra for $n = 2, 3, 4$; there were 1, 1 and 3, respectively. The two- and three-dimensional primitive parallelohedra are represented in Figures 1a and 2a. Representations of the 4-dimensional parallelohedra may be found in [10]. Here we will only be interested in the L-type of the primitive parallelohedra and will simply use the term "type" without the addition of further qualifying words.

Following Voronoĭ [2], B. N. Delone developed some new methods in the theory of parallelohedra. In the papers [13] and [14] he developed the method of empty spheres and with this method obtained in a simpler way part of the results of [2]. Then, by the method of "laminar construction", he classified the nonprimitive 4-dimensional parallelohedra [11] and together with N. N. Sandakova [15] gave another algorithm for investigating parallelohedra.

The works [16] and [17] of Minkowski as well as the articles of Venkov [18] and Coxeter [55] are devoted to the investigation of parallelohedra.

Recently, in connection with applied problems and in particular with the problem of the most economical lattice covering of n-dimensional space, the importance of enumerating all of the types of n-dimensional parallelohedra for $n > 4$ was clarified (see for example [19]–[21] and the chapter on applications (Chapter IV)). Nevertheless, attempts to apply Voronoĭ's algorithm for $n > 4$ (see for example [22] or [23]) have run into colossal computational difficulties. Even for $n = 5$ these attempts did not give the desired results: in [22] 4 types were found; in [23], completed with the help of a computer, 12 rows of 5-dimensional parallelohedra were found. However, the complete number, as is shown in our work, is equal to 221.

In connection with this, we constructed new methods for investigating the L-type for $n = 5$, the most important of which is the preliminary calculation of the C-type defined below. Some of the details of the methods proposed here have already appeared and been used to some degree by other authors. At the beginning we adopt the point of view of Voronoĭ [2] when he described his geometrical reductions (see §9 of this work). Then we take the point of view of the article of Venkov [18] in which he, for other purposes, examines domains which in fact correspond to our C-type domains. We also take into account the very interesting remarks of Dickson [24] and the dissertation of Trenerry [23].

2. We will say that two lattices belong to one and the same geometrical C-type (contiguity type)(1) if the 1-skeletons of the L-subdivision corresponding to these lattices are affinely equivalent. This definition of C-type, and also the results and structures arising from it in a natural way, complement the notion of Voronoĭ-Dirichlet domain and primitive parallelohedron.

From the definition it is clear that two lattices belonging to the same L-type belong to the same C-type, i.e. each L-type is completely contained in a single C-type.

As in the case of the L-type, we will isolate those C-types which correspond to primitive parallelohedra, i.e., we will isolate those C-types which contain at least one L-type corresponding to a primitive parallelohedron. These we will call general C-types.

Both the L-types of nonprimitive Voronoĭ-Dirichlet domains and nongeneral (special) C-types are degenerates of the general types and can be studied on this basis later. In our investigations we will not consider the special C-types as independent objects.

It is known that the L-subdivision of a primitive lattice is completely determined by the star of an arbitrarily chosen vertex. Since for a given 1-skeleton there exist only a finite number of simplicial stars, we can say that each C-type contains only a finite number of L-types.

We propose the following method for investigating the L-types of the primitive n-dimensional parallelohedra:

1. enumerate all of the C-types of lattices of a given dimension;
2. for each C-type, find the L-types contained in it.

This method, which turns out to be trivial for $n = 2$ and $n = 3$, we have carried through in its entirety for $n \leqslant 5$. We remark that both the results of Voronoĭ in the theory of parallelohedra and the achievements in this theory of a number of other researchers have been obtained using the method of continuous parameters developed by Hermite [25]. This method relates the theory of parallelohedra corresponding to Voronoĭ-Dirichlet domains to the theory of positive quadratic forms. The theory of positive quadratic forms, and in particular the reduction via perfect quadratic forms found by Voronoĭ [1], is fundamental in our work. We do not cite here the constructions and results of this theory and will assume that the reader is familiar or will make himself familiar with these through the works [1], [2] and [10]. We remark that the understanding of [1] is greatly facilitated by the commentary [26] and also the articles [27] and [28]. In addition, we assume that the reader is familiar with the general situation in the reduction theory of positive quadratic forms (for example through one of the works [10] or [29]–[31]).

3. In this work we use the following symbols:

E^n–n-dimensional Euclidean space;

(1) The term arises from the fact that for articles belonging to the same C-type the fundamental frames can be chosen so that, relative to them, the contiguity vectors of all these lattices will have the same respective coordinates. Such a choice is impossible for lattices belonging to different C-types.

n–the dimension of E^n;

$x_1, \ldots, x_n$–coordinates of a point in E^n and the variables of a quadratic form;

$f = f(x_1, \ldots, x_n) = \Sigma_{i,j=1}^n a_{ij}x_ix_j$–a quadratic form of n variables;

E^N, where $N = n(n+1)/2$–the space of coefficients of quadratic forms of n variables; the points of this space we will also call forms;

$K = K_n$–the positive (open) cone of the space E^N, i.e. the set of points corresponding to positive quadratic forms of n variables;

$\overline{K}$–the closure of K;

$\mathfrak{J}$–the group of linear transformations of E^n generated by all integral unimodular substitutions of the variables $x_1, \ldots, x_n$;

$\mathfrak{G}_1$–the group of integral unimodular linear automorphisms of Voronoĭ's reduction domain $R_1 \subset K$ with respect to the second perfect form;

$\mathfrak{G}_2$–the group of integral unimodular linear automorphisms of Voronoĭ's reduction domain $R_2 \subset K$ with respect to the third perfect form;

G, G_1 and G_2–the groups of integral unimodular substitutions of the variables $x_1, \ldots, x_n$ which generate the groups $\mathfrak{G}$, $\mathfrak{G}_1$ and $\mathfrak{G}_2$.

A linear transformation of the variables $x_1, \ldots, x_n$ which has the form $x = Sx'$, where S is some nondegenerate square matrix, we will write as $x \longrightarrow Sx$. The collection of m inequalities $a < b_1, \ldots, a < b_m$ $(a > b_1, \ldots, a > b_m)$, the left (right) sides of which are identical, we will write, using an abbreviated notation, as $a < b_1, \ldots, b_m$ $(a > b_1, \ldots, b_m)$.

Writing $i \neq j \neq \cdots \neq k$ is equivalent to asserting that all of the indices $i, j, \ldots, k$ are different.

CHAPTER I

THE C-TYPES OF n-DIMENSIONAL LATTICES: THE DERIVATION (AND ENUMERATION) OF ALL GENERAL C-TYPES FOR FOUR- AND FIVE-DIMENSIONAL LATTICES

§1. C-types: An algorithm for deducing the C-type

1. Consider two integral n-vectors; that is, two ordered collections of whole numbers $l = (l_1, \ldots, l_n)$ and $l' = (l'_1, \ldots, l'_n)$. We say that the n-vector l is congruent to the n-vector l' (modulo 2) if $l_i \equiv l'_i \pmod 2$ for $i = 1, \ldots, n$. We will call the n-vector $I = (\epsilon_1, \ldots, \epsilon_n)$, consisting of zeros and ones, where $\epsilon_1^2 + \cdots + \epsilon_n^2 > 0$, an ($n$-dimensional) index. As is easily verified, the set of indices consists of $2^n - 1$ elements. We will assume that the indices are enumerated in an arbitrary but fixed way by the numbers $1, 2, \ldots, 2^n - 1$.

It follows from [2] that in the case of primitive lattices there is a one-to-one correspondence between the indices I_σ ($\sigma = 1, \ldots, 2^n - 1$) and the edges of the L-subdivision (the contiguity vectors) in a fixed fundamental frame. Namely, each n-vector corresponding to an arbitrary contiguity vector is congruent mod 2 to some index I_σ and each index I_σ is congruent to one and only one contiguity vector. In the case of special lattices it is possible that some of the indices are congruent not to one but to two or more contiguity vectors.

DEFINITION 1.1. Let $f = f(x_1, \ldots, x_n)$ be a positive quadratic form of n variables, and I_σ an arbitrary index. The form f is said to *belong to the pretype* $l = (l_1, \ldots, l_n)$ of index I_σ, if the n-vector l (1) is congruent to the index I_σ and (2) yields a minimum of the form f in the set of all n-vectors which are congruent to I_σ. The set of all forms belonging to the pretype l of index I_σ is called the *pretype* l of that index. The collection of points of the cone K representing forms of a given pretype is called the *domain of the pretype* l *of index* I_σ, and the numbers $l_1, \ldots, l_n$ are called the *coordinates* of the pretype.

Obviously a form f belonging to the pretype l belongs also to the pretype $-l$. A form f belonging to only two pretypes, l and $-l$, of index I_σ is called a *general form* of the pretype l $(-l)$. We shall say in this case that f belongs to only one pretype of index I_σ.

Let us agree below to consider only nonempty pretypes. Obviously such pretypes belong to those and only those n-vectors l whose entries do not have a common divisor different from 1.

LEMMA 1.1. *The domain of each pretype is convex.*

The assertion follows immediately from the fact that if the vector $l = (l_1, \ldots, l_n)$ minimizes the forms f and g in the set of vectors $x = (x_1, \ldots, x_n)$, $x \equiv l \pmod 2$, then l also minimizes any of the forms on the line segment $\lambda f + (1 - \lambda)g$, $0 < \lambda < 1$.

DEFINITION 1.2. The positive form $f = f(x_1, \ldots, x_n)$ belongs to the *arithmetical C-type* defined by the collection of vectors $l_\sigma = (l_{1\sigma}, \ldots, l_{n\sigma})$, $\sigma = 1, \ldots, 2^n - 1$, if for each σ it belongs to the pretype $l_\sigma = (l_{1\sigma}, \ldots, l_{n\sigma})$ of index I_σ.

As in the case of pretype we can define arithmetical C-type and C-type domain. Obviously, the cone K decomposes into C-type domains.

DEFINITION 1.3. Two arithmetical C-types (two C-type domains) defined respectively by the collections of vectors l_σ and l'_σ $(\sigma = 1, \ldots, 2^n - 1)$ are *equivalent* if there exists an integral, unimodular, linear transformation mapping the l_σ into the l'_σ.

Keeping in mind the correspondence between the contiguity vectors of a lattice and the indices I_σ, we conclude that the geometrical C-type of a lattice agrees with the arithmetical C-type of a lattice. Namely, the geometrical C-types are equivalence classes of arithmetical C-types.

LEMMA 1.2. *The C-type domain defined by the collection of vectors* l_σ $(\sigma = 1, \ldots, 2^n - 1)$ *is the intersection of the pretype domains* $l_1, \ldots, l_{2^n-1}$ *and therefore is convex; the subdivision of the cone K into C-type domains consists of the intersection of its subdivisions into pretype domains for each of the indices* I_σ.

The proof follows quickly from the definition and Lemma 1.1.

We remark that the first definition of the L-type of a parallelohedron was given by Voronoĭ ([2], §56) and corresponds to our definition of the arithmetical L-type. The equivalence classes of the arithmetical L-types are the geometrical L-types. Below, in the case where the context clearly distinguishes between the geometrical and arithmetical L-type (C-type) we will drop these additional qualifying words.

THEOREM 1.3. *The arithmetical L- and C-types possess the following properties*:

1. *Each L-type belongs to a single C-type.*
2. *Each C-type contains a finite number of L-types.*
3. *If any two L-types are equivalent, then their corresponding C-types are equivalent.*

PROOF. Properties 1 and 3 follow directly from the fact that the definition of C-type is obtained from that of the L-type by deleting some of the conditions of the latter. Property 2 follows from arguments equivalent to those used in the Introduction to demonstrate properties of the same kind for geometrical L- and C-types.

From the well-known [2] properties of L-type domains, from Theorem 1.3 and from Lemma 1.2 we find that C-type domains possess the following properties:

1. The subdivision of K into L-type domains is a finite refinement of its subdivision into C-type domains.

2. The C-type domains are unbounded pyramids generated by convex polyhedra and have a common vertex at the origin with the cone K. They are completely contained within K.

3. The subdivision of K into C-type domains is locally finite on K.

4. In order for a form f to be an interior point of some C-type domain it is necessary and sufficient that it belong to one and only one pretype for each index.

2. DEFINITION 1.4. The set $\Omega \subset K$ is called *C-finite* (contiguity finite) if there exists a constant $\omega = \omega(\Omega)$ such that for any form $f \in \Omega$ and for any index I_σ the relation $l_1^2 + \cdots + l_n^2 < \omega$ is satisfied, where $(l_1, \ldots, l_n)$ is the pretype of the index I_σ to which the form belongs.

We will devote §2 to the criterion of C-finiteness and to the notions and general questions connected with it. Here we only remark that the domains of reduction of Minkowski [29] and Venkov [27] and the "domain $\mathcal{R}$ of reduction of Voronoĭ by perfect forms" (i.e. the collection of pairwise inequivalent perfect domains [1]) are C-finite sets.

We turn now to the description of an algorithm for determining the C-types of n-dimensional parallelohedra.

1. Corresponding to each fixed C-finite domain of reduction $\mathcal{P}$ in the cone $K \subset E^N$, there is a finite reduction algorithm. Such a domain could be, for example, the Minkowski domain, the Venkov domain or the Voronoĭ domain of reduction via perfect forms. (With respect to practical calculations our algorithm proved to be most suited to the domain of reduction via perfect forms.)

2. We subdivide the chosen domain of reduction into pretype domains (more precisely into the intersections of these pretype domains with $\mathcal{P}$). This subdivision requires a finite number of steps if $\mathcal{P}$ is a C-finite set.

3. We find the C-type domains into which $\mathcal{P}$ is subdivided. This subdivision of $\mathcal{P}$ is achieved by intersecting the subdivisions into pretype domains.

4. We select from the C-type domains which we have found those which correspond to inequivalent C-types. We remark that the process of choosing inequivalent C-types is unnecessary only in the case of domains which intersect the boundary of $\mathcal{P}$.

The set of C-types of the primitive n-dimensional parallelohedra is determined by this set of C-type domains which have been chosen.

§2. Conditions for the C-finiteness of a set and a theorem about the simple subdivision of the cone K

1. In this section we find conditions which characterize the C-finite sets $\Omega \subset K$ and with their help establish the C-finiteness of the domains mentioned in §1.2. Along with the derivation of the conditions for C-finiteness we will investigate the subdivisions of K which we have called simple. In the remaining sections of this article

we will not make use of the results of this section, and therefore it may be omitted on first reading.

DEFINITION 2.1. A quadratic form $f = f(x_1, \ldots, x_n)$ of rank r is called *simple* if it can be reduced to the form $f(x_1, \ldots, x_r)$ by an integral unimodular transformation. A bounded polyhedron $Q \subset \overline{K}$ with finitely many faces is called *simple* if all of its vertices correspond to simple quadratic forms. An unbounded pyramid with vertex the same as that of K is called a *simple pyramid* if it is generated by a simple polyhedron $Q \subset \overline{K}$.

We remark that by considering an arbitrary decomposition of a simple form into the sum of squares of linear forms we find that these forms span a subspace of the space of all linear forms which has an integer basis. From this it follows that an arbitrary linear combination of simple forms having positive coefficients is again a simple form.

The first part of the following theorem is established above in parts 1–3 of the description of the algorithm for deducing the C-types (§1). The second part is obvious.

THEOREM 2.1. *Each C-finite domain intersects a finite number of C-type domains; and, conversely, each domain $\Omega \subset K$ intersecting only a finite number of (closed) C-type domains is a C-finite domain.*

THEOREM 2.2. *In order for a set $\Omega \subset K$ to be C-finite, it is necessary and sufficient that it can be embedded in a simple pyramid. The constant ω (Definition 1.4) may be computed using this pyramid.*

For the proof of Theorem 2.2, we require some lemmas.

LEMMA 2.3. *For the C-finiteness of a set it is necessary and sufficient that it be contained in the union of a finite number of domains, each of which is equivalent to Minkowski's domain of reduction $\mathfrak{M}$.*

PROOF. The sufficiency is an obvious consequence of a theorem of Sandakova [32] (see also [15]). We prove necessity. We suppose the contrary: some C-finite set intersects an infinite number of domains equivalent to $\mathfrak{M}$. Even though it involves a slight abuse of terminology, we will call these intersections, which are of the form $\mathfrak{M} \cap \Omega$, *Minkowski domains.*

From Theorem 2.1 it follows that Ω has nonempty intersection with only a finite number of C-type domains. We will designate these intersections by Ω_j. By our hypothesis and the finiteness of the number of sets Ω_j, at least one of them must intersect with an infinite number of Minkowski domains. Transforming each of these domains into $\mathfrak{M}$ via an integer, unimodular, linear transformation, we establish the existence of points in $\mathfrak{M}$ corresponding to lattices with contiguity vectors having arbitrarily large coordinates. Since that contradicts the theorem of Sandakova, Lemma 2.3 is established.

We denote by $J(M_1, M_2)$ the "join" of the sets M_1 and M_2, i.e. the set(1) of points contained in intervals connecting points of M_1 with points of M_2. Further, we will assume that

$$J(M_1, M_2, \ldots, M_k) = J(J(M_1, M_2, \ldots, M_{k-1}), M_k),$$

where $M_1, \ldots, M_k$ are totally disconnected sets. As is easy to show, the order of the sets $M_1, \ldots, M_k$ in this definition is immaterial.

Later on, we will require positive cones and Minkowski reduction domains for forms having different numbers of variables. We will denote by K_r and $\mathfrak{M}_r$, respectively, the positive cone and Minkowski's reduction domain for the forms $f(x_{n-r+1}, \ldots, x_n)$, $r = 1, \ldots, n$. We assume the validity of the natural relations $K_1 \subset K_2 \subset \cdots \subset K_n = K$ and $\mathfrak{M}_1 \subset \mathfrak{M}_2 \subset \cdots \subset \mathfrak{M}_n = \mathfrak{M}$.

LEMMA 2.4. *Let P be the simple pyramid generated by the convex hull of the domain $\mathfrak{M}_{n-1}$ and a finite set of positive forms of n-variables (points of the cone K_n). Then for arbitrary forms $f \in P$ the inequality $|\sin f| > C > 0$ holds, where $\sin f$ is the sine of von Staudt (see, for example,* [33]) *of the form f, and $C = C(P)$ is a constant depending on P.*

PROOF. We consider Staudt's sine of the form $f = \lambda f_0 + \mu f_1$, where

$$f_0 = \sum_{i,j=1}^{n} a_{ij}x_i x_j, \quad f_1 = \sum_{s,t=n-r+1}^{n} b_{st}x_s x_t$$

are positive forms corresponding to n and r variables, $\lambda > 0$, $\mu > 0$ and $\lambda + \mu = 1$. We have

$$\sin^2 f = \frac{\det(\lambda f_0 + \mu f_1)}{\lambda a_{11} \cdots \lambda a_{n-r,\,n-r}(\lambda a_{n-r+1,\,n-r+1} + \mu b_{n-r+1,\,n-r+1}) \cdots (\lambda a_{nn} + \mu b_{nn})}$$

$$= \begin{vmatrix} a_{11} \cdots a_{1,\,n-r} & \lambda a_{1,\,n-r+1} \cdots \lambda a_{1n} \\ \cdots\cdots & \cdots\cdots \\ a_{n-r,\,1} \cdots a_{n-r,\,n-r} & \lambda a_{n-r,\,n-r+1} \cdots \lambda a_{n-r,\,n} \\ a_{n-r+1,\,1} \cdots a_{n-r+1,\,n-r} & \lambda a_{n-r+1,\,n-r+1} + \mu b_{n-r+1,\,n-r+1} \cdots \lambda a_{n-r+1,\,n} + \mu b_{n-r+1,\,n} \\ \cdots\cdots & \cdots\cdots \\ a_{n,\,1} \cdots a_{n,\,n-r} & \lambda a_{n,\,n-r+1} + \mu b_{n,\,n-r+1} \cdots \lambda a_{nn} + \mu b_{nn} \end{vmatrix}$$

$$\times [a_{11} \cdots a_{n-r,\,n-r}(\lambda a_{n-r+1,\,n-r+1} + \mu b_{n-r+1,\,n-r+1}) \cdots (\lambda a_{nn} + \mu b_{nn})]^{-1}. \tag{2.1}$$

Consider λ to be very small. Then from (2.1) it follows that

$$\sin^2 f \approx \sin^2 f^* \sin^2 f_1, \tag{2.2}$$

(1) We remark that we have considered the case where M_1 and M_2 are convex sets in euclidean space. In this case $J(M_1, M_2)$ coincides with the convex hull of $M_1 \cup M_2$. However, we will not make use of this fact in our development.

where $f^*(x_1, \ldots, x_{n-r}) \equiv f_0(x_1, \ldots, x_{n-r}, 0, \ldots, 0)$ is a positive form of $n-r$ variables. As is easily shown,

$$\sin^2 f^* \geqslant \sin^2 f_0. \tag{2.3}$$

On the basis of (2.2) and (2.3) we deduce the existence of neighborhoods $Of_0 \subset K_n$ and $Of_1 \subset K_r$ of the points f_0 and f_1, respectively, and also the existence of numbers $0 < \lambda_0 < 1$ and $0 < \alpha_0 < 1$ such that

$$\sin^2(\lambda f + \mu g) > \alpha_0 \sin^2 f_0 \sin^2 f_1, \tag{2.4}$$

when $f \in Of_0$, $g \in Of_1$ and $0 \leqslant \lambda \leqslant \lambda_0$.

Let $F_0 \subset K_n$ and $F_1 \subset K_r$ be compact sets. Then on the strength of the Heine-Borel lemma and (2.4) there are numbers $\alpha_1 > 0$, $\lambda_1 > 0$ and a neighborhood $O'f_1 \subset K_r$ such that for all $0 \leqslant \lambda \leqslant \lambda_1$ and for arbitrary forms $f \in F_0$ and $g \in O'f_1$ the inequality

$$\sin^2(\lambda f + \mu g) > \alpha_1 \sin^2 f_1 \tag{2.5}$$

is fullfilled.

Applying again the Heine-Borel lemma, but this time to the set F_1, we find $\alpha_2 > 0$ and $\lambda_2 > 0$ such that the inequality

$$\sin^2(\lambda f + \mu g) > \alpha_2 \tag{2.6}$$

holds for all $0 \leqslant \lambda \leqslant \lambda_2$ and arbitrary forms $f \in F_0$ and $g \in F_1$.

For all $\lambda_2 \leqslant \lambda \leqslant 1$ and for arbitrary $f \in F_0$ and $g \in F_1$ the set of forms of the form $\lambda f + \mu g$ lies in the interior of K_n and is compact. Thus there is a number $\alpha_3 > 0$ such that for an arbitrary form from this set the inequality

$$\sin^2(\lambda f + \mu g) > \alpha_3 \tag{2.7}$$

holds. From the combination of (2.6) and (2.7) we get

$$\sin^2(\lambda f + \mu g) > \min(\alpha_2, \alpha_3) = \alpha > 0$$

for all $f \in F_0$, $g \in F_1$, $\lambda \geqslant 0$ and $\mu \geqslant 0$, $\lambda + \mu = 1$, i.e., we have

$$\sin^2 f > \alpha > 0, \tag{2.8}$$

for $f \in J(F_0, F_1)$.

Let $n > r_1 > \cdots > r_q$. Let $F_0 \subset K_n$ and $F_1 \subset K_{r_1}, \ldots, F_q \subset K_{r_q}$ be compact sets, and let $f \in J(F_0, \ldots, F_q)$. We will show the existence of an $\alpha > 0$

for which the forms f satisfy (2.8).

We suppose that our assertion is correct for all $f \in J_i = J(F_0, \ldots, F_i)$ for $i < q$ and show that it holds for $i = q$. We may represent a positive form $f \in J_q$ as $f = \lambda f_1 + \mu f_2$, where $f_1 \in J_{q-1}$ and $f_2 \subset F'_q$.

Denoting $f^*(x_1, \ldots, x_{n-r_{q-1}}) \equiv f_1(x_1, \ldots, x_{n-r_{q-1}}, 0, \ldots, 0)$, we have, by the induction hypothesis and by the inequality analogous to (2.3), that $\sin^2 f^* > \sin^2 f_1 > \alpha$ for $f_1 \in J_{q-1}$. Further, since the coefficients of the forms $f \in J_{q-1}$ are bounded, a sufficiently small $\lambda_0 > 0$ may be chosen such that for all forms g in a sufficiently small neighborhood $Of_2 \subset K_{r_q}$ (positive!), where $f_2 \in K_{r_q}$, and for all forms $f \in J_{q-1}$, the relationship $\sin(\lambda f + \mu g) > ½\alpha \sin f_2$ holds for all $0 \leqslant \lambda \leqslant \lambda_0$.

Employing the Heine-Borel lemma as above, we obtain

$$\sin^2(\lambda f_1 + \mu f_2) > \beta_1, \tag{2.9}$$

where $\beta_1 > 0$ is a fixed number. Let U_1 denote the set of forms of the form $\lambda f_1 + \mu f_2$ which satisfy (2.9).

We now choose a compact set $F'_{q-1} \subset K_{r_{q-1}}$ and an open set $O_{q-1} \subset F'_{q-1}$ such that $J(F_{q-1}, F_q) \subset U_1 \cup O_{q-1}$. By assumption our assertion holds for all forms from the set $J(F_0, \ldots, F_{q-2}, F'_{q-1})$. Therefore, for arbitrary forms $f \in U_2 = J(F_0, \ldots, F_{q-2}, F'_{q-1}) \cup U_1$ we have the inequality $\sin^2 f > \beta_2 > 0$, where β_2 is a constant.

Again we choose a compact set $F'_{q-2} \subset K_{r_{q-2}}$ and an open set $O_{q-2} \subset F'_{q-2}$ such that $J(F_{q-2}, F_{q-1}, F_q) \subset U_2 \cup O_{q-2}$, and repeat the previous argument. It follows that if $f \in U_3 = J(F_0, \ldots, F_{q-3}, F'_{q-2}) \cup U_2$, then $\sin^2 f > \beta_3 > 0$, where β_3 is some constant. Repeating this argument q times, we find that (2.8) holds for an arbitrary form $f \in J_q$.

From the homogeneity of Staudt's sine it follows that (2.8) holds for arbitrary forms belonging to the pyramid generated by the set J_q. The assertion of the lemma follows from the fact that the simple pyramid under consideration may be represented in the form

$$J(F_n, F_{n-1}, \ldots, F_0) = J(F_0, F_1, \ldots, F_n),$$

where F_i $(1 \leqslant i \leqslant n)$ is the convex hull of the union of the nondegenerate forms of determinant 1 contained in the domain $\mathfrak{M}_{n-i+1}$ and F_0 is the convex envelope of the finite number of nondegenerate forms (of n variables) mentioned in the statement of the lemma.

LEMMA 2.5. *Let P be the simple pyramid generated by the complex hull of the domain $\mathfrak{M}_{n-1}$ and a finite set of positive forms $\varphi_1, \ldots, \varphi_m \subset K_n$. Then there exists a constant $C > 0$ such that for arbitrary i, j $(i, j = 1, \ldots, n)$ and for an arbitrary form $f \in P$*

$$\left|\frac{a_{ij}}{a_{ii}}\right| < C. \tag{2.10}$$

PROOF. We represent the form $f \in P$ as $f = \lambda f_0 + \mu f_1$, where f_0 is a form contained in the convex hull Q of the points $\varphi_1, \ldots, \varphi_m$, f_1 is contained in $\mathfrak{M}_{n-1}$, and $\lambda \geqslant 0$, $\mu \geqslant 0$, and $\lambda + \mu = 1$. Since Q is compact, the relation $|a^0_{ij}/a^0_{ii}| < C'$ holds for the form f_0, where $C' > 0$ is some constant. According to [29] (see also [15], [32], and [33]) an analogous inequality holds for f_1, i.e. $|a^1_{ij}/a^1_{ii}| < C''$, for some constant C''.

We choose $C = \max(C', C'')$ and obtain

$$\left|\frac{a_{ij}}{a_{ii}}\right| = \left|\frac{\lambda a^0_{ij} + \mu a^1_{ij}}{\lambda a^0_{ii} + \mu a^1_{ii}}\right| \leqslant \frac{\lambda\,|a^0_{ij}| + \mu\,|a^1_{ij}|}{\lambda a^0_{ii} + \mu a^1_{ii}} < \frac{\lambda C' a^0_{ii} + \mu C'' a^1_{ii}}{\lambda a^0_{ii} + \mu a^1_{ii}} \leqslant C,$$

and so finish the proof of the lemma.

LEMMA 2.6. *Each simple pyramid is generated by a finite number of Minkowski domains.*

PROOF. Consider an arbitrary N-dimensional simple pyramid P, and the bounded $(N-1)$-dimensional polyhedron P^* which generates it. We construct the barycentric subdivision of P^* and then consider the closed $(N-1)$-dimensional simplices P^*_i $(i = 1, \ldots, s)$ of the refinement and the simple pyramids generated by them.

Obviously, it is sufficient to carry the proof through for each of the simple pyramids P_i. First we point out that an arbitrary pyramid P_i has only a single face Q_i which is not shared with any other simple pyramid. The face Q_i is, in turn, a simple pyramid in a positive cone of a smaller dimension.

We introduce the induction hypothesis that our lemma holds for all dimensions $n < n_0$; for $n = 1$ and 2 the lemma is obvious, and we consider it to be known. According to the induction hypothesis, the face Q_i is covered by a finite number of Minkowski domains of the corresponding dimension.

Thus, P_i is covered by the union of the convex hulls of its nondegenerate edges and a finite number of Minkowski domains of smaller dimension. By Lemmas 2.4 and 2.5, P_i is covered by a finite number of domains of Sandakova's type [33], i.e. the pyramid P_i is C-finite. By Lemma 2.3, P_i is covered by a finite number of Minkowski domains.

PROOF OF THEOREM 2.2. Let the set Ω be embedded in the simple pyramid P. Then by Lemma 2.6 it is covered by a finite number of Minkowski domains. Therefore on the strength of Lemma 2.3 it is C-finite. The calculation of the constant $\omega(P)$ is carried through by calculating this constant for each of the Minkowski domains individually. For estimates of this constant see [15], [32] and [33].

Assume that the set Ω is C-finite. Then by Lemma 2.3 it is covered by a finite number of Minkowski domains. Obviously, the convex hull of these domains is a simple pyramid. This finishes the proof.

2. DEFINITION 2.2. A subdivision of the cone $\overline{K}$ into simple pyramids will be called *simple* if it is invariant with respect to the group $\mathfrak{G}$ and if it has only a finite number of elements which are mutually inequivalent with respect to $\mathfrak{G}$.

Examples of simple subdivisions are: 1) the subdivision of the cone $\overline{K}$ into domains which are equivalent to the reduction domain of Minkowski [29] or the reduction domain of Venkov [27], 2) the perfect subdivision [1], or 3) the subdivision into L- or C-type domains.

THEOREM 2.7. *Let P_1 and P_2 be two arbitrary simple subdivisions of the cone $\overline{K}$. Then each of the elements of the first subdivision intersects only a finite number of elements of the second subdivision, and conversely, i.e., the class of simple subdivisions is closed with respect to finite intersections.*

PROOF. By Lemma 2.2 and Theorem 2.2 we conclude that each element of either of the subdivisions P_1 or P_2 is covered by a finite number of Minkowski domains. Further, as in the method of proof of Lemma 2.6, we can show that each Minkowski domain intersects with no more than a finite number of domains from either subdivision P_1 or P_2.

Take an arbitrary element from P_1. It is covered by a finite number of Minkowski domains. And since each of these intersects only a finite number of elements of P_2, we can conclude that each element of P_1 intersects at most a finite number of elements of P_2. This finishes the proof.

§3. Analysis of C-types on the perfect domains R and R_1. The general C-types for $n = 2, 3$

1. We will apply the algorithm for deducing the C-types which was described in §1. For the reduction domain P we will take the reduction domain by perfect forms, R. For a description of R see [1], or [28].

It is known [1] that for arbitrary $n \geqslant 2$ the domain R contains the domain R of the first perfect form. The domain R consists of the set of quadratic forms of the form

$$f(x_1, x_2, \ldots, x_n) = \sum_{i=1}^{n} \rho_i x_i^2 + \sum_{\substack{i,\, j=1 \\ (i<j)}}^{n} \rho_{ij} (x_i - x_j)^2; \qquad \rho_i, \rho_{ij} \geqslant 0. \tag{3.1}$$

It is not difficult to show that for an arbitrary index I_σ ($\sigma = 1, \ldots, 2^n - 1$), the points of R belong to the pretype domain I_σ. From this it follows that R is entirely contained within a single C-type domain. Since it is well known ([2], §104) that R consists of a single L-type domain, we could have bypassed this argument.

If we consider a face of R, we discover that the pretype of any fixed index of an arbitrary point of the face is not uniquely defined. For example, each point of the face $\rho_{12} = 0$, opposite to the edge $\lambda(x_1 - x_2)^2$, belongs to the pretype $I = (1, 1, 0, \dots, 0)$ of index I as well as to the pretype $(1, -1, 0, \dots, 0)$ of the same index. All of the faces of R are mutually equivalent.

From the above it follows that R is a C-type domain, where $l_{i\sigma} = \epsilon_{i\sigma}$ $(i = 1, \dots, n; \sigma = 1, \dots, 2^n - 1)$. We remark that it follows from a theorem of Dickson [24] that R is the unique perfect domain which is coincident with an L-type domain. Furthermore, for $n = 2$ or 3, R coincides with $\mathcal{R}$. This establishes the fact that there is one and only one general C-type in the spaces E^2 and E^3. The coincidence of the C- and L-subdivision shows that in these spaces the 1-skeleton of the primitive L-subdivision uniquely determines the L-subdivision.

2. Starting with $n = 4$, $\mathcal{R}$ contains a domain R_1 which is the domain of the second perfect form. R_1 consists of the set of forms of the form

$$\begin{aligned} f(x_1, x_2, \dots, x_n) = \sum_{i=1}^{n} a_i x_i^2 + \sum_{k=3}^{n} [b_{1k}(x_1 - x_k)^2 \\ + b_{2k}(x_2 - x_k)^2 + c_k(x_1 + x_2 - x_k)^2] \\ + \sum_{\substack{k, m=3 \\ (k<m)}}^{n} [c_{km}(x_k - x_m)^2 + d_{km}(x_1 + x_2 - x_k - x_m)^2], \end{aligned} \tag{3.2}$$

where $a_i, b_{1k}, b_{2k}, c_k, c_{km}, d_{km} \geqslant 0$.

Under Voronoĭ's transformation ([1], §36)

$$\begin{aligned} x_1 &\to x_1 + x_2, \\ x_q &\to x_1 - x_q \qquad (q = 2, 3, \dots, n), \end{aligned} \tag{3.3}$$

the forms of R_1 appear as

$$f(x_1, x_2, \dots, x_n) = \sum_{\substack{i, j=1 \\ (i<j)}}^{n} [\rho_{ij}(x_i + x_j)^2 + \rho_{\overline{ij}}(x_i - x_j)^2], \tag{3.4}$$

where $\rho_{ij}, \rho_{\overline{ij}} \geqslant 0$.

Since the determinant of the transformation (3.3) is 2, this transformation induces a change to a coordinate frame which for each of the lattices is not basic. In the new coordinate frame corresponding to the forms (3.4), the points of the lattice consist of either integer vectors or vectors which can be written as the sum of an integer vector and the vector $(\frac{1}{2}, \dots, \frac{1}{2})$.

The coefficients ρ_{ij} and $\rho_{\overline{ij}}$ of (3.4) uniquely determine the coefficients of the quadratic form $f = \sum_{i,j=1}^{n} a_{ij} x_i x_j$. Namely:

$$a_{ii} = \sum_{\substack{j=1 \\ (j \neq i)}}^{n} (\rho_{ij} + \rho_{\overline{ij}}); \qquad a_{ij} = \rho_{ij} - \rho_{\overline{ij}} \qquad (i, j = 1, 2, \dots, n; \ i < j). \tag{3.6}$$

For this reason ρ_{ij} and $\rho_{\overline{ij}}$, except for the requirement $\rho_{ij}, \rho_{\overline{ij}} \geqslant 0$, may be considered as a parametrization of the entire space E^N. However, the nonuniqueness of this representation (3.4) of forms of R_1 adds a further complication, and special care must be exercised in the investigation of these forms.

For example, a linear dependency among the parameters ρ_{ij} and $\rho_{\overline{ij}}$ does not guarantee that an $(N-1)$-dimensional subspace in E^N is defined. Obviously, a linear dependency will define an $(N-1)$-dimensional subspace only in the case when it is equivalent to some linear dependency among the coefficients a_{ij} of the quadratic form. That is, the linear dependency among the variables ρ_{ij} and $\rho_{\overline{ij}}$, which is expressed by setting a linear form equal to zero, must be expressible in a similar manner in terms of the a_{ij}. This comment is important for the following example.

Voronoĭ observed [1] that among the faces of the domain R_1 are faces which we will denote by F_ν and D_μ and will describe in accordance with [28]. Each of the faces F_ν $(\nu = 1, \ldots, n \cdot 2^{n-1})$ is determined by the conditions

$$\rho'_{1i} = \rho'_{2i} = \cdots = \rho'_{i-1,i} = \rho'_{i,i+1} = \cdots = \rho'_{in} = 0, \tag{3.6}$$

where i has one of the values $1, \ldots, n$ and ρ'_{ij} is either ρ_{ij} or $\rho_{\overline{ij}}$. (Additional meaning will be attached to the coefficients ρ'_{ij} below.) Each of the faces D_μ $(\mu = 1, \ldots, n \cdot 2^{n(n-1)/2})$ is determined by $n(n-1)/2$ relations

$$\rho'_{jk} = 0 \qquad (j,\ k = 1,\ 2, \ldots, n;\ \ j \neq k \neq i;\ \ j < k), \tag{3.7}$$

where $i = 1, \ldots, n$ is fixed. Conversely, each of the conditions (3.6) and (3.7) determines one of the faces F_ν or D_μ.

If the parameters ρ_{ij} and $\rho_{\overline{ij}}$ are nonnegative, conditions (3.6) and (3.7) are respectively equivalent to the following linear relations:

$$\sum_{\substack{j=1 \\ (j \neq i)}}^{n} \rho'_{ij} = 0, \tag{3.8}$$

$$\sum_{\substack{j,\ k=1 \\ (j \neq k \neq i;\ j<k)}}^{n} \rho'_{jk} = 0. \tag{3.9}$$

If negative values are allowed for ρ_{ij} and $\rho_{\overline{ij}}$, then each of the linear relations (3.8) and (3.9) determines an $(N-1)$-dimensional plane in E^N, and the corresponding inequalities determine half-spaces.

PROPOSITION 3.1. *The linear relations* (3.8) *and* (3.9) *define* $(N-1)$-*dimensional planes in* E^N, *and the inequalities*

$$\sum_{\substack{j=1\\(j\neq i)}}^{n} \rho'_{ij} > 0, \qquad \sum_{\substack{j,k=1\\(j\neq k\neq i;\ j<k)}}^{n} \rho'_{jk} > 0 \qquad (i=1,\ 2,\ \ldots,\ n)$$

determine half-spaces both of which contain R_1.

PROOF. On the basis of (3.5) we have

$$2\sum_{\substack{j=1\\(j\neq i)}}^{n} \rho'_{ij} = a_{ii} - \sum_{\substack{j=1\\(j\neq i)}}^{n} \varepsilon_{ij} a_{ij},$$

$$2\sum_{\substack{j,k=1\\(j\neq k\neq i;\ j<k)}}^{n} \rho'_{jk} = \sum_{m=1}^{n} a_{mm} - 2\left(a_{ii} + \sum_{\substack{j,k=1\\(j\neq k\neq i;\ j<k)}}^{n} \varepsilon_{jk} a_{jk}\right)$$

$$(i=1,\ 2,\ \ldots,\ n),$$

where ϵ_{ij} is equal to $+1$ if in the left-hand side of the identity we have $\rho'_{ij} = \rho_{ij}$, and to -1 if $\rho'_{ij} = \rho_{\overline{ij}}$. Since we have represented each of the linear relations (3.8) and (3.9) between the parameters ρ_{ij} and $\rho_{\overline{ij}}$ as a linear relation among the coefficients a_{ij} of a quadratic form, the proposition is proved.

In the sequel we will make use of negative values of the parameters ρ_{ij} and $\rho_{\overline{ij}}$, as well as the fact that (3.4) applies to all forms $f \in R_1$ and $\rho_{ij}, \rho_{\overline{ij}} \geqslant 0$ in this case. In this latter case we will say that the form under consideration has a nonnegative decomposition according to formula (3.4).

3. The transformation (3.3) translates the set of indices I_σ ($\sigma = 1, \ldots, 2^{n-1}$) into the set of indices J_σ, among which $2^n - 1$ indices have the form

$$J^d_q = (\delta_{1q}, \delta_{2q}, \ldots, \delta_{nq}), \quad \delta_{iq} = \pm\frac{1}{2}; \quad i=1,\ 2,\ \ldots,\ n;$$
$$q=1,\ 2,\ \ldots,\ 2^{n-1}, \tag{3.10}$$

and the remaining $2^{n-1} - 1$ indices have the form

$$J^c_p = (\delta_{1p}, \delta_{2p}, \ldots, \delta_{np}), \quad \delta_{ip} = 0,\ 1; \quad i=1,\ 2,\ \ldots,\ n;$$
$$p=1,\ 2,\ \ldots,\ 2^{n-1}-1, \tag{3.11}$$

It is assumed that the case $\delta_{1p} = \delta_{2p} = \cdots = \delta_{np} = 0$ does not occur. The total number of entries equalling $-\tfrac12$ in (3.10) and the total number of entries equalling 1 in (3.11) may not exceed $n/2$. In addition, for even n, if there are $n/2$ entries equalling $-\tfrac12$ in (3.20), then $\delta_{1q} = \tfrac12$, and if there are $n/2$ entries equalling 1 in (3.11), then $\delta_{1p} = 1$.

In contrast to the index I_σ, we will say that the coordinate vector $l = (l_1, \ldots, l_n)$ is congruent to the index $J_\sigma = (\epsilon_{1\sigma}, \ldots, \epsilon_{n\sigma})$ if the congruence

$$l_1 - \varepsilon_{1\sigma} \equiv l_2 - \varepsilon_{2\sigma} \equiv \ldots \equiv l_n - \varepsilon_{n\sigma} \pmod 2 \tag{3.12}$$

is satisfied. When (3.12) is satisfied we write $l \equiv J_\sigma \pmod 2$.

THEOREM 3.2. *Forms $f \in R_1$ belong to the pretypes of index J_q^d ($q = 1, \ldots, 2^{n-1}$), where J_q^d is of the form* (3.10). *If $f \notin \bigcup_\nu F_\nu$, then it belongs to no other pretype of the index J_q^d. If $f \in \bigcup_\nu F_\nu$, then f belongs to two or more pretypes which correspond to one of the indices J_q^p.*

PROOF. Take an arbitrary point $f \in R_1$, $f \in \bigcup_\nu F_\nu$, and consider the nonnegative expansion of (3.4). We will consider some index J_q^d and a coordinate vector $l = (\delta_{1q} + l_1, \ldots, \delta_{nq} + l_n)$ which is congruent to this index, where $l_1, \ldots, l_n$ are whole numbers and $l \neq \pm J_q^d$.

On the basis of (3.4) we find that

$$f(l) - f(J_q^d) = \sum_{\substack{i,j=1\\(i\neq j)}}^{n} \{\rho_{ij}[(\delta_{iq} + l_i + \delta_{jq} + l_j)^2 - (\delta_{iq} + \delta_{jq})^2] \\ + \rho_{\overline{ij}}[(\delta_{iq} + l_i - \delta_{jq} - l_j)^2 - (\delta_{iq} - \delta_{jq})^2]\}. \tag{3.13}$$

Let the numbers $l_1, \ldots, l_n$ be even; then some $l_{i_0} = 2m$, where $|m| \geqslant 1$. In this case, by considering the first part of (3.13) we get

$$f(l) - f(J_q^d) \geqslant 4 \sum_{\substack{j=1\\(j\neq i_0)}}^{n} \rho'_{i_0 j}, \tag{3.14}$$

where again $\rho'_{i_0 j}$ is either $\rho_{i_0 j}$ or $\rho_{\overline{i_0 j}}$; the latter depends on the relationship of the signs of numbers l_{i_0} and l_j and on which of the two numbers $\delta_{iq} + \delta_{jq}$ or $\delta_{iq} - \delta_{jq}$ is equal to zero. Inequality (3.14) reduces to an equality, for example, in the case of a pretype l where $l_{i_0} = -4\delta_{i_0 q}$ and $l_j = 0$ ($j = 1, \ldots, n; j \neq i_0$).

If the numbers $l_1, \ldots, l_n$ are odd, then either we have that $l_{i_0} + \delta_{i_0 q} \neq -\delta_{i_0 q}$ for some index i_0, or else $l = -J_q^d$, a case which is excluded. And in this case, as is easy to see, (3.14) holds. When $l_{i_0} = 2\delta_{i_0 q}$ and $l_j = -2\delta_{jq}$ ($j = 1, \ldots, n$; $j \neq i_0$), (3.14) becomes an equality.

It follows from the nonnegativity of the coefficients ρ_{ij} and $\rho_{\overline{ij}}$ that the right-hand side of (3.14) (independently of the choice of decomposition of f by (3.4)) is strictly positive for points of R_1 belonging to the pretypes $l \equiv J_q^d \pmod 2$, $l \neq \pm J_q^d$; the right-hand side may equal zero at points of the faces F_ν corresponding to some of these pretypes. With this the theorem is proved.

From the set of faces $\{D_\mu\}$ defined by (3.7) or by (3.9), we isolate the subset of faces H_ν ($\nu = 1, \ldots, n \cdot 2^{n-1}$) at which R_1 adjoins domains which are equivalent to R.

Let $(\omega_1, \ldots, \omega_n)$ be an ordered collection of numbers of which one, say ω_i, is equal to zero, the remaining being equal to either $+1$ or -1. There are $n \cdot 2^{n-1}$ members of this family.

We may describe the faces H_ν after establishing the following one-to-one correspondence between these faces and the set of n-tuples $\pm(\omega_1, \ldots, \omega_n)$. The face H corresponding to the n-tuple $(\omega_1, \ldots, \omega_n)$ is the set of all those and only those points of R_1 for which the relation

$$\sum_{\substack{j,k=1 \\ (j\neq k\neq i;\ j<k)}}^{n} \rho'_{jk} = 0, \tag{3.15}$$

is satisfied, where $\rho'_{jk} = \rho_{jk}$ if $\omega_j\omega_k = 1$ and $\rho'_{jk} = \rho_{\overline{jk}}$ if $\omega_j\omega_k = -1$.

THEOREM 3.3. *Corresponding to each of the n indices* $(1, 0, \ldots, 0)$, $(0, 1, \ldots, 0)$, $\ldots$, $(0, 0, \ldots, 1)$ *of the form* (3.11), *forms* $f \in R_1$ *belong to pretypes which coincide with this index; if* $f \notin (\bigcup_\nu H_\nu) \cup (\bigcup_\nu F_\nu)$, *then it cannot belong to any other pretypes corresponding to these indices. If* $f \in \bigcup_\nu H_\nu$, *then it belongs to two or more pretypes congruent to at least one of the above-mentioned indices.*

PROOF. Obviously, without loss of generality, it is sufficient to carry the proof through for any one of the indices named in the theorem, for example, the index $(1, 0, \ldots, 0)$. As in Theorem 3.2, we assume that f has a nonnegative expansion of the form (3.4).

Suppose that $l = (1 + l_1, l_2, \ldots, l_n) \neq \pm(1, 0, \ldots, 0)$ is congruent to this index. According to (3.4) we have

$$f(l) - f(1, 0, \ldots, 0) = \sum_{j=2}^{n} \{\rho_{1j}[(1 + l_1 + l_j)^2 - 1] + \rho_{\overline{1j}}[(1 + l_1 - l_j)^2 - 1]\} + \sum_{\substack{i,j=2 \\ (i<j)}}^{n} [\rho_{ij}(l_i + l_j)^2 + \rho_{\overline{ij}}(l_i - l_j)^2]. \tag{3.16}$$

For arbitrary odd numbers $l_1, \ldots, l_n$ we have

$$f(l) - f(1, 0, \ldots, 0) \geqslant 4 \sum_{\substack{i,j=2 \\ (i<j)}}^{n} \rho'_{ij}, \tag{3.17}$$

where $\rho'_{ij} = \rho_{ij}$ if $l_il_j > 0$, and $\rho'_{ij} = \rho_{\overline{ij}}$ if $l_il_j < 0$. When $l_1, \ldots, l_n$ are even, one of the following two inequalities is satisfied:

$$f(l) - f(1, 0, \ldots, 0) \geqslant 8 \sum_{j=2}^{n} (\rho_{1j} + \rho_{\overline{1j}}), \tag{3.18}$$

if $l_2 = \cdots = l_n = 0$, or

$$f(l) - f(1, 0, \ldots, 0) \geqslant 4 \sum_{\substack{j=2 \\ (j\neq i_0)}}^{n} \rho'_{i_0j} + 8\rho'_{1i_0} \geqslant 4 \sum_{\substack{j=1 \\ (j\neq i_0)}} \rho'_{i_0j}, \tag{3.19}$$

where $|l_{i_0}| > 0$, and i_0 has one of the values $2, 3, \ldots, n$.

As is easy to see, the right side of (3.18) never vanishes at points of R_1, since we have assumed that the cone K is open. The right side of (3.19) vanishes only at certain points $f \in \bigcup_\nu F_\nu$. From a comparison of equations (3.15) for the faces H_ν and inequality (3.17) we obtain that the right side of (3.17) vanishes precisely when f belongs to some face H_ν. The first statement of the theorem is proved.

Let $(0, \omega_2, \ldots, \omega_n)$ be the n-tuple mentioned above corresponding to H_ν. Then all points of H_ν belonging to the pretype $l^0 = (-1, \omega_2, \ldots, \omega_n)$ satisfy the equality $f(l^0) = f(1, 0, \ldots, 0)$. In a similar way, for any face H_ν given in advance we can indicate an index and at least two pretypes to which each point of H_ν belongs relative to this index. This establishes the second assertion of the theorem.

We remark that since the faces $\{H_\nu\}$ are generally faces of domains which are equivalent to R as well as faces of R_1, the second statement of Theorem 3.3 may be established from the properties of R mentioned at the beginning of this section.

§4. The general C-types for $n = 4$

1. From Theorems 3.2 and 3.3 it follows that for forms $f(x_1, x_2, x_3, x_4) \in R_1$ twelve of the fifteen pretypes have already been considered, leaving only those pretypes corresponding to the indices

$$J_{12} = (1, 1, 0, 0), \qquad J_{13} = (1, 0, 1, 0), \qquad J_{14} = (1, 0, 0, 1).$$

We denote by l^{ij} $(i, j = 1, 2, 3, 4; i \neq j)$ a coordinate vector for which $l_i = l_j = 1$ and the remaining coordinates are zero. The coordinate vector $l^{\bar{i}\bar{j}}$ distinguishes itself from l^{ij} only in that one of its coordinates, l_i or l_j, is equal to -1.

LEMMA 4.1. *If the quadratic form $f \in R_1$ of four variables does not lie on a face of R_1 of dimension less than* 9, *then, corresponding to each of the indices J_{1p}* $(p = 2, 3, 4)$, *the form f belongs to one or more of the following four pretypes*

$$l^{1p},\ l^{\bar{1}\bar{p}},\ l^{qr},\ l^{\bar{q}\bar{r}} \qquad (p, q, r = 2, 3, 4;\ p \neq q \neq r)$$

and does not belong to any others having these indices. The membership of the form in one or another of these pretypes is determined by the smallest of the four parameters $\rho_{1p}, \rho_{\bar{1}\bar{p}}, \rho_{qr}$ or $\rho_{\bar{q}\bar{r}}$, and the smallest of these is independent of the representation of f in the form (3.4).

PROOF. We remark that each of the inequalities which distinguishes the smallest of the parameters $\rho_{1p}, \rho_{\bar{1}\bar{p}}, \rho_{qr}$ or $\rho_{\bar{q}\bar{r}}$ defines a half-space of E^N $(N = 10)$. In fact, from (3.5) we have

$$\rho_{1p} - \rho_{\bar{1}\bar{p}} = a_{1p}, \qquad \rho_{qr} - \rho_{\bar{q}\bar{r}} = a_{qr},$$
$$4(\rho_{1p} - \rho_{qr}) = a_{11} + a_{pp} - a_{qq} - a_{rr} + 2(a_{1p} - a_{qr}),$$

which suffices for the verification of this remark.

The first assertion of the lemma is not concerned with the representation of the form in the form (3.4), and consequently does not depend on the choice of representation. If the second statement holds for any representation of f in the form (3.4), it holds for an arbitrary representation, and consequently this statement also does not depend on the choice of representation. It follows that we can prove the lemma if we can demonstrate its truth for each form $f \in R_1$ in some fixed representation of the form (3.4). Below we will assume that such a choice of representation has been made, and moreover that the selected decomposition is nonnegative.

Since a permutation of the coordinate indices permutes the indices J_{12}, J_{13}, J_{14} and does not vary the form of the quadratic form (3.4), i.e. leaves R_1 invariant, it is sufficient to prove the lemma for only one of the indices, for example J_{12}.

After substituting the coordinate vectors $l^{12} = (1, 1, 0, 0)$, $l^{\overline{12}} = (1, -1, 0, 0)$, $l^{34} = (0, 0, 1, 1)$, $l^{\overline{34}} = (0, 0, 1, -1)$ into (3.4) we obtain

$$\begin{aligned} f(l^{12}) &= 4\rho_{12} + \eta, & f(l^{\overline{12}}) &= 4\rho_{\overline{12}} + \eta, \\ f(l^{34}) &= 4\rho_{34} + \eta, & f(l^{\overline{34}}) &= 4\rho_{\overline{34}} + \eta, \end{aligned} \tag{4.1}$$

where $\eta = \rho_{13} + \rho_{14} + \rho_{23} + \rho_{24} + \rho_{\overline{13}} + \rho_{\overline{14}} + \rho_{\overline{23}} + \rho_{\overline{24}}$.

From (4.1) we see that if there are one or several minima among the numbers $\rho_{12}, \rho_{\overline{12}}, \rho_{34}$ and $\rho_{\overline{34}}$, then f will belong correspondingly to one or several of the pretypes l^{12}, $l^{\overline{12}}$, l^{34} or $l^{\overline{34}}$, provided there is no other coordinate vector l congruent to the index J_{12} giving f a lower value.

We will now show that such a coordinate vector $l = (l_1, l_2, l_3, l_4)$ cannot exist. We remark that $l_1 \equiv l_2 \pmod 2$, $l_3 \equiv l_4 \pmod 2$ and $l_1, l_2 \not\equiv l_3, l_4 \pmod 2$. Consequently, every coordinate vector $l \equiv J_{12} \pmod 2$ differing from those considered either has fewer than two zeros or, if it has two zeros, then it also has a coordinate whose modulus is not less than three. Let us suppose that $l_1 > 0$, $l_2 < 0$, $l_1 \equiv l_2 \equiv 1 \pmod 2$ and $l_1 \geqslant |l_2|$.

Comparing $f(l)$ with $f(l^{\overline{12}})$, we get

$$f(l) - f(l^{\overline{12}}) \geqslant \begin{cases} 4\rho_{12} + 12\rho_{\overline{12}} + 8(\rho_{13} + \rho_{\overline{13}} + \rho_{14} + \rho_{\overline{14}}), \\ \qquad \text{if } l_3 + l_4 = 0, \ l_1 > |l_2|; \\ 32\rho_{12} + 8 \sum\limits_{s=1,\,2;\ t=3,\,4} (\rho_{st} + \rho_{\overline{st}}), \\ \qquad \text{if } l_3 = l_4 = 0, \ l_1 = |l_2|; \\ 8(\rho'_{13} + \rho'_{23}) + 4(\rho_{34} + \rho_{\overline{34}}), \\ \qquad \text{if } l_3 \neq 0, \ l_4 = 0; \\ 8(\rho'_{13} + \rho'_{23} + \rho'_{14} + \rho'_{24}) + 16\rho'_{34}, \\ \qquad \text{if } l_3 \neq 0, \ l_4 \neq 0. \end{cases} \tag{4.2}$$

The right-hand side of (4.2) (again we mention that the right-hand side is independent of the representation of f in the form (3.4)) is strictly positive at all interior points of R_1 and at all interior points of the 9-dimensional faces on its boundary (see [1], [28], (4.5) and (4.6)). Only on faces of dimension less than or equal to 8 can the right-hand side of (4.2) be equal to zero.

As is easy to see, the above situation holds for a general $l \equiv J_{12} \pmod 2$ which is not equal to one of the pretypes l^{12}, $l^{\overline{12}}$, l^{34} or $l^{\overline{34}}$; i.e., for such an l the inequality $f(l) - f(\widetilde{l}) > 0$ is satisfied at all points interior to R_1 and at all points interior to any of the 9-dimensional faces on its boundary. Here $\widetilde{l}$ is one of the pretypes l^{12}, $l^{\overline{12}}$, l^{34} or $l^{\overline{34}}$. This finishes the proof of the lemma.

Three out of the four parameters of any of the three quartets

$$\rho_{1p}, \rho_{\overline{1p}}, \rho_{qr}, \rho_{\overline{qr}} \quad (p, q, r = 2, 3, 4; \quad p \neq q \neq r), \tag{4.3}$$

which were mentioned in Lemma 4.1 may be taken to be independent and free of relations with any of the other quartets. This follows since none of the parameters of a form $f \in R_1$ occur simultaneously in two such quartets. From this and from Lemma 4.1 it follows that R_1 decomposes into $4^3 = 64$ C-type domains. (We use the term "C-type domain of R_1" for the intersection of a C-type domain with the perfect domain R_1.) Each of these domains is characterized by three parameters which correspond to the smallest parameters belonging to each of the three quartets of (4.3). For example, the three parameters $\rho_{12}, \rho_{13}, \rho_{14}$ correspond to the C-type domain characterized by the inequalities

$$\rho_{12} < \rho_{\overline{12}}, \rho_{34}, \rho_{\overline{34}};$$
$$\rho_{13} < \rho_{\overline{13}}, \rho_{24}, \rho_{\overline{24}};$$
$$\rho_{14} < \rho_{\overline{14}}, \rho_{23}, \rho_{\overline{23}}.$$

2. We will choose from the 64 C-type domains of R_1 a collection of domains which are mutually inequivalent with respect to the group $\mathfrak{G}_1$ of automorphisms of R_1. The group $\mathfrak{G}_1$ induces the following transformations of variables x_1, x_2, x_3, x_4 in (3.4): 1) an arbitrary permutation of a number of the variables; 2) a change of sign of an arbitrary variable; 3) the transformation

$$x_i \to \sum_{j=1}^{4} \frac{1}{2} x_j - x_i \quad (i = 1, \ldots, 4).$$

It follows that $\mathfrak{G}_1$ induces the following transformations of the parameters ρ_{ij} and $\rho_{\overline{ij}}$ $(i, j = 1, \ldots, 4; i < j)$ of (3.4): 1) a permutation of the indices; 2) a change of sign of an arbitrary index (a change of sign, for example, of index i denotes a transformation under which $\rho_{ij} \longrightarrow \rho_{\overline{ij}}$ and $\rho_{\overline{ij}} \longrightarrow \rho_{ij}$ $(j = 1, \ldots, 4; j \neq i)$, the remaining parameters remaining fixed); 3) the transformation (for $n = 4$ only!)

$$\rho_{\overline{ij}} \to \rho_{\overline{ij}}, \quad \rho_{ij} \to \rho_{km} \quad (i, j, k, m = 1, 2, 3, 4; \quad i \neq j \neq k \neq m). \tag{4.4}$$

For ease in selecting the mutually inequivalent domains we introduce a diagram for each of the C-type domains. We number the vertices of a square with the numbers

1, 2, 3, 4 consecutively, and trace three segments with ends at these vertices. The numbers of the ends of each of these segments we take to be those of the double indices of the three parameters which define the given domain. In this context the parameter ρ_{ij} would correspond to the line segment with ends at the vertices i and j drawn with a solid line (black line), and the parameter $\rho_{\overline{ij}}$ would correspond to a segment with the same ends, but traced with a dotted line (red line).

It is not difficult to verify that the figures corresponding to the C-type domains are of two types, "triangular" and "forked", both consisting of black and red segments. (We remark that forks correspond to the faces F_ν.) The above description of the transformations of $\mathfrak{G}_1$, a permutation of the indices or a change of sign of an arbitrary number of parameters, implies the equivalence of all of the forks, and also implies that among the triangles there are two which are mutually inequivalent. One of these types of triangles either has all black edges of has two red and one black edge. We will call it "black triangular". The second type of triangle has either all red edges or two black and one red edge. We will call it "red triangular". A black triangle corresponds to a C-type domain having faces of the form H_ν.

With the help of transformation (4.4) we can show that "forks" are equivalent to black triangles, i.e. for $n = 4$ the faces F_ν and H_ν are equivalent.

Having established that none of the transformations of $\mathfrak{G}_1$ transforms black triangles (or any of the forks) into red triangles, we have the following lemma:

LEMMA 4.2. *The perfect domain $R_1 \subset K_4$ intersects two C-type domains, inequivalent under the group $\mathfrak{G}_1$, namely Δ_1 ($\Delta_1 \cap R_1 = \{\rho_{12}, \rho_{13}, \rho_{14}\}$) and Δ_2 ($\Delta_2 \cap R_1 = \{\rho_{\overline{12}}, \rho_{\overline{13}}, \rho_{\overline{23}}\}$).*

In the case $n = 4$, according to [1] and [28], the domain R_1 is bounded by faces of two types: 1) faces at which R_1 adjoins domains which are equivalent to R—all of these faces are mutually equivalent under $\mathfrak{G}_1$ and are equivalent, for example, to the following face:

$$\rho_{12} + \rho_{13} + \rho_{14} = 0; \tag{4.5}$$

2) faces at which R_1 adjoins domains which are equivalent to it. These faces are also mutually equivalent under the group $\mathfrak{G}_1$. A representative of these faces is given by (2)

$$\rho_{\overline{12}} + \rho_{\overline{13}} + \rho_{\overline{23}} = 0. \tag{4.6}$$

From the fact that R and the domains equivalent to it in the cone K are C-type domains, or from Theorems 3.2 and 3.3, it follows that the set $\{\rho_{12}, \rho_{13}, \rho_{14}\}$ comprises an entire C-type domain Δ_1. In fact none of the 9-dimensional faces of the domain $\{\rho_{12}, \rho_{13}, \rho_{14}\}$ is equivalent to the face (4.6), and

(2) Since for $n = 4$ the faces F_ν and H_ν ($\nu = 1, \ldots, 32$) of the perfect domain R_1 are equivalent under $\mathfrak{G}_1$, the faces H_ν are not formally represented among the faces (4.5) and (4.6).

the pretypes of index $(-\frac{1}{2}, \frac{1}{2}, \frac{1}{2}, \frac{1}{2})$ are not constant on the face (4.5).

The intersection $\{\rho_{\overline{12}}, \rho_{\overline{13}}, \rho_{\overline{23}}\}$ of the domain Δ_2 with R_1 consists of the 9-dimensional face (4.6). The 12 pretypes which are constant for all of the forms of R_1 (Theorems 3.2 and 3.3) are constant on this face. According to Lemma 4.1, the remaining three pretypes of forms from the set $\{\rho_{\overline{12}}, \rho_{\overline{13}}, \rho_{\overline{23}}\}$ are constant on the face (4.6). Thus the C-type domain Δ_2 whose intersection with R_1 consists of the set $\{\rho_{\overline{12}}, \rho_{\overline{13}}, \rho_{\overline{23}}\}$ extends into the domain R_1' which is equivalent to R_1 and is contiguous to it at the face (4.6).

In addition, it is now clear that the C-type domains Δ_1 and Δ_2 are inequivalent not only under the group $\mathfrak{G}_1$ but also under the group $\mathfrak{G}$, since only one of them lies entirely in R_1. We may now summarize our results concerning the investigation of the C-types for $n = 4$ in the following theorem:

THEOREM 4.3. *In the space E^4 there are three general C-type parallelohedra. The domain Δ_0 of the first coincides with the domain R. The domain Δ_1 of the second is entirely contained in the domain R_1. The domain Δ_2 of the third is situated partly in R_1 and partly in the domain R_1' which is equivalent to R_1.*

From the theorem it follows that for $n = 4$ the subdivision of the cone K into C-type domains, in contradistinction to the cases $n = 2, 3$, is not a refinement of the perfect subdivision.

Comparison of the C-type domains with the known [2] L-type domains shows that the C-type domain Δ_1 coincides with one of the type II domains of Voronoĭ, and the intersection $\{\rho_{\overline{12}}, \rho_{\overline{13}}, \rho_{\overline{23}}\}$ of the third-C-type domain Δ_2 with R_1 yields the type III domain of Voronoĭ. This last C-type domain thus decomposes into L-type domains. Below (§9) this decomposition will be investigated in detail. It turns out to be sufficiently clear now that the domain of the third C-type separates (at the face (4.6) of R_1) into two L-type domains which are mutually equivalent and consequently equivalent to the type III domain of Voronoĭ.

§5. Diagrams of the C-type domains of the domain R_1 ($n = 5$)

1. For $n = 5$ there are three perfect forms and correspondingly three perfect domains, R, R_1 and R_2 (see [1]). We have already discussed R and R_1 in §3 for arbitrary n. In that section the treatment of the C-types of the domain R was complete, but that for R_1 was only partial (Theorems 3.2 and 3.3).

We now describe the construction which allowed us to find the C-type domains intersecting R_1. Our method of finding these domains and isolating from them a set of mutually inequivalent domains is a generalization of that used in §4 for the case $n = 4$. As in §3, we will employ the change of variables (3.3) and assume that the forms of R_1 are expressed as in (3.4).

The case $n = 5$ requires 31 indices. From Theorems 3.2 and 3.3 it follows that for 21 of the indices the pretypes of R_1 are uniquely determined (with the exception of the forms belonging to the parts of the boundary mentioned in the statement of the theorem); that is, for each of these indices, R_1 contains only a single pretype. The question regarding the subdivision into pretype domains for the remaining 10 indices is left. These indices are the following:

$$J_{ij} = (\varepsilon_1, \ldots, \varepsilon_5) \qquad (i, j = 1, 2, \ldots, 5;\ i \neq j),$$

where $\epsilon_i = \epsilon_j = 1$, the remaining coordinates of the index being equal to zero. Obviously $J_{ij} = J_{ji}$.

Let i, j, k, m, p be a permutation of the numbers 1, 2, 3, 4, 5 in which $k < m < p$. We examine the six pretypes of index J_{ij} having the following designations and coordinates:

$$\begin{array}{lll} l^{ij} & : \ l_i = l_j = 1, & l_k = l_m = l_p = 0; \\ l^{\overline{ij}} & : \ l_i = -l_j = 1\ (i<j), & l =_k l_m = l_p = 0; \\ l^{0,\,ij} & : \ l_i = l_j = 0, & l_k = l_m = l_p = 1; \\ l^{1,\,ij} & : \ l_i = l_j = 0, & -l_k = l_m = l_p = 1; \\ l^{2,\,ij} & : \ l_i = l_j = 0, & l_k = -l_m = l_p = 1; \\ l^{3,\,ij} & : \ l_i = l_j = 0, & l_k = l_m = -l_p = 1. \end{array} \tag{5.1}$$

LEMMA 5.1. *Corresponding to each of the indices* J_{ij} $(i, j = 1, \ldots, 5; i < j)$ *the domain* R_1 *is subdivided into the six pretype domains* (5.1). *Whether a form* $f \in R_1$ *belongs to one or another of these pretypes depends on which of the six quantities*

$$\rho_{ij},\ \rho_{\overline{ij}},\ \rho_{km} + \rho_{kp} + \rho_{mp},\ \rho_{\overline{km}} + \rho_{\overline{kp}} + \rho_{mp},\ \rho_{\overline{km}} + \rho_{kp} + \rho_{\overline{mp}},\ \rho_{km} + \rho_{\overline{kp}} + \rho_{\overline{mp}}$$
$$(i, j, k, m, p = 1, \ldots, 5;\ i \neq j \neq k \neq m \neq p;\ k < m < p) \tag{5.2}$$

is smallest, and that property is invariant with respect to the representation of the quadratic form in the form (3.4) (*the order of enumeration of the quantities* (5.2) *corresponds to that of the pretypes in the list* (5.1)).

If the form f is an interior point of the domain R_1 *or an interior point of any of its 14-dimensional faces, then it does not belong to any pretype of index* J_{ij} *other than those listed in* (5.1).

PROOF. By using the relations

$$\rho_{ij} - \rho_{\overline{ij}} = a_{ij},$$

$$4(\rho_{km} + \rho_{kp} + \rho_{mp} - \rho_{ij})$$
$$= a_{kk} + a_{mm} + a_{pp} - a_{ii} - a_{jj} + 2(a_{km} + a_{kp} + a_{mp} - a_{ij})$$

obtained from (3.5) it is easy to show that each of the inequalities which arises by comparing the smallest of the list (5.2) with the remaining defines a half-space of E^N $(N = 15)$, or, in the case of an equality, a 14-dimensional plane.

Regarding the invariance of the choice of the least among the quantities (5.2) with respect to the representation of the form in the form (3.4), we only mention that this has been considered in the proof of Lemma 4.1. We assume some fixed nonnegative decomposition as in (3.4) for each form $f \in R_1$.

Without losing generality, it is sufficient to carry through the proof of the lemma for any one of the indices, for example J_{12}. Substituting the coordinates of each of the pretypes (4.1) of index J_{12} into (3.4), we get

$$
\begin{aligned}
f(l^{12}) &= 4\rho_{12} + \sigma,\\
f(l^{0,12}) &= 4(\rho_{34} + \rho_{35} + \rho_{45}) + \sigma,\\
f(l^{2,12}) &= 4(\rho_{\overline{34}} + \rho_{35} + \rho_{\overline{45}}) + \sigma,\\
f(l^{\overline{12}}) &= 4\rho_{\overline{12}} + \sigma,\\
f(l^{1,12}) &= 4(\rho_{\overline{34}} + \rho_{\overline{35}} + \rho_{45}) + \sigma,\\
f(l^{3,12}) &= 4(\rho_{34} + \rho_{\overline{35}} + \rho_{\overline{45}}) + \sigma,
\end{aligned}
\tag{5.3}
$$

where $\sigma = \Sigma_{r=3}^{5} (\rho_{1r} + \rho_{2r} + \rho_{\overline{1r}} + \rho_{\overline{2r}})$.

From (5.3) it can be seen at once that the membership of the form f in one or more of the pretypes (5.1) is determined by those quantities among the six quantities (5.2) which are minimal. By virtue of (5.3) the invariance of the selection of the smallest among (5.2) mentioned in the lemma is obvious.

We will now show that forms $f \in R_1$, besides (5.1), belong to no other pretypes of index J_{12} if we exclude the case where f belongs to a face of R_1 of dimension less than 14.

Let $l = (l_1, \ldots, l_5) \equiv J_{12} \pmod 2$ be distinct from any of the pretypes (5.1). If both of the coordinates l_1 and l_2 are nonzero, we will compare it either with the pretype l^{12} or with $l^{\overline{12}}$, depending on whether the signs of l_1 and l_2 are equal or different. If only one of the coordinates l_1 or l_2 of the pretype l is equal to zero, then all three coordinates l_3, l_4, l_5 are nonzero. In that case we compare it with that pretype $l^{q,12}$ ($q = 0, 1, 2, 3$) where agreement with respect to the signs of the coordinates l_3, l_4 and l_5 is achieved.

We consider the difference $f(l) - f(\widetilde{l})$, where $\widetilde{l}$ is the pretype of the list (5.1) which we must compare to the pretype l according to the rule formulated above. Since the investigations are analogous for all six possibilities, we restrict ourselves to the detailed consideration of only one of them.

Let, for example, $\widetilde{l} = l^{12}$. For definiteness, take $l_1 \geqslant l_2$ and $|l_3| \geqslant |l_4| \geqslant |l_5|$; then in the case of odd l_1 and l_2 we have

$$
f(l) - f(l^{12}) \geqslant \begin{cases} 12\rho_{12} + 4\rho_{\overline{12}} + 8 \sum\limits_{r=3}^{5} (\rho_{1r} + \rho_{\overline{1r}}), \\ \qquad \text{if } l_1 > l_2; \\ 32\rho_{12} + 8 \sum\limits_{r=3}^{5} (\rho_{1r} + \rho_{2r} + \rho_{\overline{1r}} + \rho_{\overline{2r}})_r \\ \qquad \text{if } l_1 = l_2, \;\; l_3 = l_4 = l_5 = 0; \\ 8(\rho'_{13} + \rho'_{23}) + 4(\rho_{34} + \rho_{35} + \rho_{\overline{34}} + \rho_{\overline{35}}), \\ \qquad \text{if } l_1 = l_2, \;\; l_3 \neq 0, \;\; l_4 = l_5 = 0; \\ 4 \sum\limits_{r=3}^{4} [2(\rho'_{1r} + \rho'_{2r}) + \rho_{r5} + \rho_{\overline{r5}}] + 16\rho'_{34}, \\ \qquad \text{if } l_1 = l_2, \;\; l_3 \neq 0, \;\; l_4 \neq 0, \;\; l_5 = 0; \\ 8 \sum\limits_{r=3}^{5} (\rho'_{1r} + \rho'_{2r}) + 16(\rho'_{34} + \rho'_{35} + \rho'_{45}), \\ \qquad \text{if } l_1 = l_2, \;\; l_3 \neq 0, \;\; l_4 \neq 0, \;\; l_5 \neq 0. \end{cases} \tag{5.4}
$$

Under the assumption of even l_1 and l_2 we have

$$
f(l) - f(l^{12}) \geqslant 12\rho_{12} + 8 \sum_{r=3}^{5} (\rho'_{1r} + \rho'_{2r}) + \rho'_{34} + \rho'_{35} + \rho'_{45}. \tag{5.5}
$$

In all cases the right-hand sides of (5.4) and (5.5) are strictly positive, except for those forms f belonging to the intersection of at least two faces of the form F_ν (see (3.8)).

Since there is an analogous situation for all six cases $\widetilde{l}$, we may assume that Lemma 5.1 has been proved.

2. From Lemma 5.1 it follows that the domain R_1 intersects not more than 6^{10} C-type domains, since each of these domains is defined by $6 \cdot 9 = 54$ inequalities. We denote by M this set of 6^{10} intersections. In order to determine the C-types of R_1 it is necessary to clarify which of the elements μ of M are domains, i.e. are not empty and have interior points; then we must select from M a mutually inequivalent set. In order to make our resolution of this question completely transparent, we establish a correspondence between the elements of M and some geometrical diagrams to be defined below.

As a preliminary, let us introduce for the sets $\mu \in M$ the following notation. Corresponding to the pretypes of index J_{ij} of the list (5.1), and in accordance with the order of this list, we introduce the letters and numbers a, b, 0, 1, 2, 3. Next, assuming that the 10 indices J_{ij} ($i, j = 1, \ldots, 5$; $i < j$) are disposed in lexicographical order, we write in this order, for each of the indices, the notation for the pretypes of (5.1) uniquely defining a set μ. Equivalently, a set μ will be denoted by an ordered string of 10 symbols, which we will call a symbolic string. For example, the symbolic string *ba0aa3aabb* denotes the domain defined by the pretypes $l^{\overline{12}}$, l^{13}, $l^{0,14}$, l^{15}, l^{23}, $l^{3,24}$, l^{25}, l^{34}, $l^{\overline{35}}$, $l^{\overline{45}}$ and consequently the inequalities

$$
\begin{array}{lllll}
\rho_{\overline{12}} < \rho_{12}, & \rho_{34} + \rho_{35} + \rho_{45}, & \rho_{\overline{34}} + \rho_{\overline{35}} + \rho_{45}, & \rho_{\overline{34}} + \rho_{35} + \rho_{\overline{45}}, & \rho_{34} + \rho_{\overline{35}} + \rho_{\overline{45}}; \\
\rho_{13} < \rho_{\overline{13}}, & \rho_{24} + \rho_{25} + \rho_{45}, & \rho_{\overline{24}} + \rho_{\overline{25}} + \rho_{45}, & \rho_{\overline{24}} + \rho_{25} + \rho_{\overline{45}}, & \rho_{24} + \rho_{\overline{25}} + \rho_{\overline{45}}; \\
\rho_{23} + \rho_{25} + \rho_{35} < \rho_{14}, & \rho_{\overline{14}}, & \rho_{\overline{23}} + \rho_{\overline{25}} + \rho_{35}, & \rho_{\overline{23}} + \rho_{25} + \rho_{\overline{35}}, & \rho_{23} + \rho_{\overline{25}} + \rho_{\overline{35}}; \\
\rho_{15} < \rho_{\overline{15}}, & \rho_{23} + \rho_{24} + \rho_{34}, & \rho_{\overline{23}} + \rho_{\overline{24}} + \rho_{34}, & \rho_{\overline{23}} + \rho_{24} + \rho_{\overline{34}}, & \rho_{23} + \rho_{\overline{24}} + \rho_{\overline{34}}; \\
\rho_{23} < \rho_{\overline{23}}, & \rho_{14} + \rho_{15} + \rho_{45}, & \rho_{\overline{14}} + \rho_{\overline{15}} + \rho_{45}, & \rho_{\overline{14}} + \rho_{15} + \rho_{\overline{45}}, & \rho_{14} + \rho_{\overline{15}} + \rho_{\overline{45}}; \\
\rho_{13} + \rho_{\overline{15}} + \rho_{\overline{35}} < \rho_{24}, & \rho_{\overline{24}}, & \rho_{13} + \rho_{15} + \rho_{35}, & \rho_{\overline{13}} + \rho_{\overline{15}} + \rho_{35}, & \rho_{\overline{13}} + \rho_{15} + \rho_{\overline{35}}; \\
\rho_{25} < \rho_{\overline{25}}, & \rho_{13} + \rho_{14} + \rho_{34}, & \rho_{\overline{13}} + \rho_{\overline{14}} + \rho_{34}, & \rho_{\overline{13}} + \rho_{14} + \rho_{\overline{34}}, & \rho_{13} + \rho_{\overline{14}} + \rho_{\overline{34}}; \\
\rho_{34} < \rho_{\overline{34}}, & \rho_{12} + \rho_{15} + \rho_{25}, & \rho_{\overline{12}} + \rho_{\overline{15}} + \rho_{25}, & \rho_{\overline{12}} + \rho_{15} + \rho_{\overline{25}}, & \rho_{12} + \rho_{\overline{15}} + \rho_{\overline{25}}; \\
\rho_{\overline{35}} < \rho_{35}, & \rho_{12} + \rho_{14} + \rho_{24}, & \rho_{\overline{12}} + \rho_{\overline{14}} + \rho_{24}, & \rho_{\overline{12}} + \rho_{14} + \rho_{\overline{24}}, & \rho_{12} + \rho_{\overline{14}} + \rho_{\overline{24}}; \\
\rho_{\overline{45}} < \rho_{45}, & \rho_{12} + \rho_{13} + \rho_{23}, & \rho_{\overline{12}} + \rho_{\overline{13}} + \rho_{23}, & \rho_{\overline{12}} + \rho_{13} + \rho_{\overline{23}}, & \rho_{12} + \rho_{\overline{13}} + \rho_{\overline{23}}.
\end{array}
\tag{5.6}
$$

We understand that some of the 6^{10} symbolic strings may correspond to empty sets or sets of dimension less than $N = 15$.

In order to construct diagrams corresponding to each of the sets $\mu \in M$, we label the vertices of a regular pentagon either with the numbers 1, 2, 3, 4, 5 or with the letters *i, j, k, m, p* which denote a permutation of 1, 2, 3, 4, 5. Besides the vertices, a diagram will contain some segments connecting the vertices, which are traced either with solid lines (black edges) or dotted lines (red edges). The diagrams will also contain what we call *b*-triangles. Each *b*-triangle *kmp* is obtained by applying a homothetic transformation of coefficient $0 < \lambda < 1$ to the triangle with vertices *k, m, p* and is situated inside it (see the diagrams of Table I in §7, below). The sides of a *b*-triangle are traced with either solid or dotted lines and are called correspondingly black or red sides of the *b*-triangle.

Corresponding to the pretypes (5.1) of index J_{ij} we will associate to our diagrams the following elements:

l^{ij}—the black segment *ij*;

$l^{\overline{ij}}$ —the red segment *ij*;

$l^{0,ij}$—the *b*-triangle *kmp* with black sides;

$l^{r,ij}$—the *b*-triangle *kmp* with two red sides, both ending for $k < m < p$ respectively at the vertex k for $r = 1$, m for $r = 2$, and p for $r = 3$, and the third side black.

Each set $\mu \in M$ corresponds to a diagram consisting of 10 elements, red and black edges and *b*-triangles. The choice of these elements is determined by the pretypes (5.1) of each index J_{ij}, which define the set μ. For example, the set *ba0aa3aabb* corresponds to diagram 40 of Table I.

In this context, to each pair *i, j* of vertices of the diagram of the set μ there corresponds either a black segment *ij*, or a red segment *ij*, or any one of four *b*-triangles *kmp* depending on the given μ. Thus if the diagram has the *b*-triangle *kmp*, then it does not have the edge *ij*, and conversely.

Obviously, if the diagram of a set contained in R_1 is given, just as with the symbolic string, the system of inequalities which determine this set may be constructed in a unique fashion.

3. We will now consider which of the sets $\mu \in M$ are nonempty and have interior points, i.e. the question of nontrivial intersections of C-type domains with R_1. We recall that ρ'_{ij} $(i, j = 1, \ldots, 5; i \neq j)$ designates all of the coefficients of (3.4), ρ_{ij} or $\rho_{\overline{ij}}$, and that i, j, k, m, p designates some permutation of 1, 2, 3, 4, 5.

We will call the quantities (5.2) the *weights* of the corresponding pretype of index J_{ij}; we call the weights ρ_{ij} and $\rho_{\overline{ij}}$ *simple*, the remaining, *composite.*

In the diagrams of the sets $\mu \in M$ the simple weights correspond to black and red edges and the composite to b-triangles.

We agree to say that the diagram of the set μ *contains the triangle kmp* (let us emphasize, not a b-triangle) if it contains the three edges km, kp and mp of this triangle.

LEMMA 5.2. *If the form $f \in R_1$ has a pretype of index J_{ij} of composite weight and is an interior point of a C-type domain, then its pretypes corresponding to the indices J_{km}, J_{kp} and J_{mp} have simple weights.*

PROOF. Let us assume the contrary. The form f has a composite weight of index J_{ij} and (for example) of index J_{km}. Then, among others, the following inequalities hold:

$$\begin{gathered} \rho'_{km}+\rho'_{kp}+\rho'_{mp}<\rho_{ij}, \quad \rho'_{km}+\rho'_{kp}+\rho'_{mp}<\rho_{\overline{ij}}, \\ \rho'_{ij}+\rho'_{ip}+\rho'_{jp}<\rho_{km}, \quad \rho'_{ij}+\rho'_{ip}+\rho'_{jp}<\rho_{\overline{km}}. \end{gathered} \tag{5.7}$$

From (5.7) we have

$$\rho'_{km}<\rho_{ij}, \quad \rho'_{km}<\rho_{\overline{ij}}, \quad \rho'_{ij}<\rho_{km}, \quad \rho'_{ij}<\rho_{\overline{km}}. \tag{5.8}$$

The inconsistency of the system (5.8) establishes the lemma.

We will require the following reformulation of Lemma 5.2 in the investigation of the diagrams of the sets μ.

LEMMA 5.3. *If the set μ is not empty and has an interior point, and if its diagram contains a b-triangle kmp, then it contains the triangle kmp consisting of the edges km, kp and mp.*

From Lemma 5.2 Corollaries 1–3 enumerated below easily follow. Regarding the formulation of these corollaries we assume that $f \in R_1$ is an interior point of a C-type domain.

COROLLARY 1. *An arbitrary form has no more than 4 indices whose pretypes have composite weights. If these are four such indices, they are* J_{ij}, J_{ik}, J_{im} *and* J_{ip}.

COROLLARY 2. *Each triple of indices whose pretypes have composite weights are either triples of the form* J_{ij}, J_{ik}, J_{im} *or triples of the form* J_{ij}, J_{ik}, J_{jk}.

COROLLARY 3. *Every pair of indices whose pretypes have composite weights differs by no more than indexing from the pair of indices* J_{ij}, J_{ik}.

In this section we will call a set a *C-type domain* if it is either a C-type domain lying entirely in R_1 or the intersection of a C-type domain with R_1 having an interior point in R_1.

With respect to the diagrams of the sets μ Corollaries 1–3 are equivalent to the following lemma:

LEMMA 5.4. *If* $\mu \in M$ *is a C-type domain, then its diagram has no more than four b-triangles. If the diagram contains* 4 *or* 2 *b-triangles, then they must arrange themselves with respect to one another in a unique way, as depicted in Figures* 5a *and* 5b, *respectively. If there are three b-triangles in the diagram, then there are two possible arrangements* (*Figures* 5c *and* 5d).

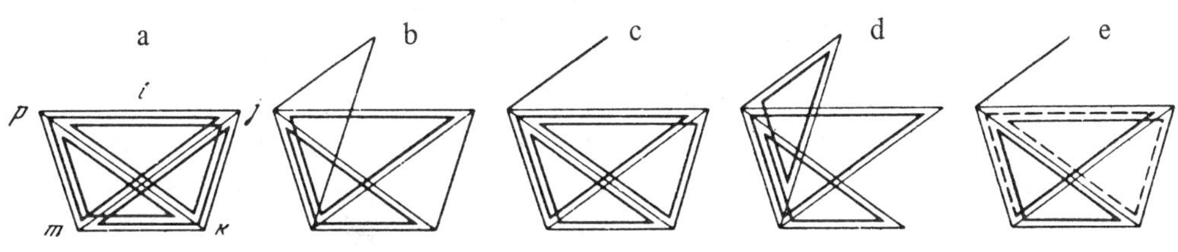

FIGURE 5

LEMMA 5.5. *Let an interior point f of a C-type domain* ($f \in R_1$) *belong to a pretype of index* J_{ij} *of composite weight* $\rho'_{km} + \rho'_{kp} + \rho'_{mp}$. *Then, among the three pretypes of index* J_{km}, J_{kp}, J_{mp}, *at least two have weights from the set* $\rho'_{km}, \rho'_{kp}, \rho'_{mp}$.

PROOF. According to Lemma 5.2, the pretypes of index J_{km}, J_{kp} and J_{mp} have simple weights. We denote by ρ''_{qr} (not ρ'_{qr}) the second simple weight of the pretype of index J_{qr}. Assuming that the pretypes of index J_{km} and J_{kp} have respectively the weights ρ''_{km} and ρ''_{kp}, we get the inequalities $\rho''_{km} < \rho'_{km}$ and $\rho''_{kp} < \rho'_{kp}$, from which it follows that $\rho''_{km} + \rho''_{kp} + \rho'_{mp} < \rho'_{km} + \rho'_{kp} + \rho'_{mp}$. The last inequality contradicts the fact that the pretype of index J_{ij} has weight $\rho'_{km} + \rho'_{kp} + \rho'_{mp}$. The lemma is proved.

LEMMA 5.6. *In the diagram of a C-type domain at most two of the sides of a b-triangle kmp may differ in color from the corresponding edges of the triangle kmp.*

LEMMA 5.7. *If an interior point f of a C-type domain* ($f \in R_1$) *belongs to the pretypes of index* J_{ij}, J_{ik}, J_{im} *with composite weights, then these weights may not be the three*

$$\rho_{km}+\rho_{kp}+\rho_{mp},\ \rho_{jm}+\rho_{\overline{jp}}+\rho_{\overline{mp}},\ \rho_{\overline{jk}}+\rho_{jp}+\rho_{\overline{kp}}. \tag{5.9}$$

PROOF. If the pretypes of the indices of the lemma were the three with the weights (5.9), then we would have, among others, the three inequalities

$$\rho_{km}+\rho_{kp}+\rho_{mp}<\rho_{km}+\rho_{\overline{kp}}+\rho_{\overline{mp}},\quad \rho_{jm}+\rho_{\overline{jp}}+\rho_{\overline{mp}}<\rho_{jm}+\rho_{jp}+\rho_{mp},$$
$$\rho_{\overline{jk}}+\rho_{jp}+\rho_{\overline{kp}}<\rho_{\overline{jk}}+\rho_{\overline{jp}}+\rho_{kp}.$$

From the sum of these inequalities we get a contradictory inequality which establishes the lemma.

In the language of diagrams, Lemma 5.7 may be formulated in the following way:

LEMMA 5.8. *If a diagram of a C-type domain contains three b-triangles disposed respectively as in Figure* 5c, *then the colours of the sides of these b-triangles may not be as in* 5e.

From the analysis of Lemmas 5.2–5.8 and their corollaries it follows that only a few of the 6^{10} sets $\mu \in \mathsf{M}$ are C-type domains. Lemmas 5.2–5.8 give criteria as to which sets $\mu \in \mathsf{M}$ are trivially empty or do not have interior points.

We denote by $\mathsf{M}_0 \subset \mathsf{M}$ the totality of all sets μ which according to Lemmas 52.–5.8 may have interior points. As will be obvious from further investigations, each of the sets $\mu \in \mathsf{M}_0$ is a C-type domain. For that reason we will now make use of this name.

4. We will now show how a subset of elements mutually inequivalent with respect to the group $\mathfrak{G}_1$ may be selected from the set M_0.

For $n = 5$ the group $\mathfrak{G}_1$ (see [1], §§35 and 36, and [28]) for the given domain R_1 of forms is defined by the following transformations of the parameters ρ_{ij} and $\rho_{\overline{ij}}$ ($i, j = 1, \ldots, 5$; $i < j$): 1) permutations of the indices; 2) a change in sign of any of the indices (a change in sign of index i denotes the transformation $\rho_{ij} \to \rho_{\overline{ij}}$, $\rho_{\overline{ij}} \to \rho_{ij}$, $\rho_{jk} \to \rho_{jk}$, $\rho_{\overline{jk}} \to \rho_{\overline{jk}}$ ($j, k = 1, \ldots, 5$; $j \neq k \neq i$; $j < k$)).

Obviously, the transformation "change the sign of index i", in terms of the diagram of the domain μ, recolors the edges and sides of b-triangles which terminate at the vertex i of the pentagon: black edges and sides go over to red, and conversely. We will call this transformation of the diagram "a recoloring at the ith vertex".

We have the following observation:

PROPOSITION 5.9. *Two domains μ_1 and μ_2 are inequivalent with respect to the group $\mathfrak{G}_1$ if and only if there does not exist any renumbering or recoloring of the diagrams of these domains under which the diagrams coincide.*

Since in the investigation of mutual inequivalence we will consider the diagrams of sets $\mu \in \bar{M}_0$ rather than the corresponding C-type domains, Proposition 5.9 along with Lemmas 5.3, 5.4, 5.6 and 5.8 will lie at the base of all further considerations in the following section.

We point out some features in the diagrams of domains $\mu \in \bar{M}_0$ which are invariant with respect to the group $\mathfrak{G}_1$. The appearance of a triangle (b-triangle) occurs in one of two ways: as a black triangle which can be transformed into a triangle with all edges black by a recoloring, or as a red triangle which can be transformed into a triangle with all edges red by a recoloring. From the rule associating pretypes to b-triangles it follows that all of the b-triangles of these diagrams are black.

The second invariant feature relates to a set of six edges belonging to four of the vertices of the diagram. Let us agree to call such a configuration a tetrahedron. With elementary testing we find that all tetrahedrons are of one of three types: 1) black, in which all four of the triangles are black; 2) red, in which all of the triangles are red; 3) mixed, in which two triangles are black and two are red.

§6. Derivation of the C-types of the five-dimensional parallelohedra of the domain R_1

1. In this section we concern ourselves with the derivation of the C-types. However, we consider not the general equivalence of these domains but rather only equivalence with respect to $\mathfrak{G}_1$. In fact, the 81 C-type domains found here which are mutually inequivalent with respect to $\mathfrak{G}_1$ do not correspond to 81 general C-type five-dimensional parallelohedra; in §8 we discover that they correspond to only 75.

Obviously, the number of pretypes corresponding to composite weights of the given C-type domain is invariant with respect to the group $\mathfrak{G}_1$. We will carry the discussion of mutual inequivalence through considering separately domains characterized by zero, one, two, three or four composite weights (0-, 1-, 2-, 3- and 4- composite types).

In the case of the 0-composite C-types the diagrams do not have a single b-triangle, and consequently all 10 edges are traced.

In order to find all of the diagrams of the mutually inequivalent domains of 0-composite type we append four edges with ends at the vertex 5 to one of the three tetrahedrons described above with vertices at the points 1, 2, 3, 4. Black tetrahedrons are composed of all black edges. There are only three inequivalent possibilities for selecting four edges with ends at vertex 5 in the case of a black tetrahedron. These are represented by: 1) all black edges; 2) three black edges; and 3) two black edges. Equivalently, starting with a black tetrahedron we get three C-type diagrams (Figure 6a, b, c). Analogous considerations starting with a red tetrahedron yield three C-type diagrams (Figure 6d, e, f).

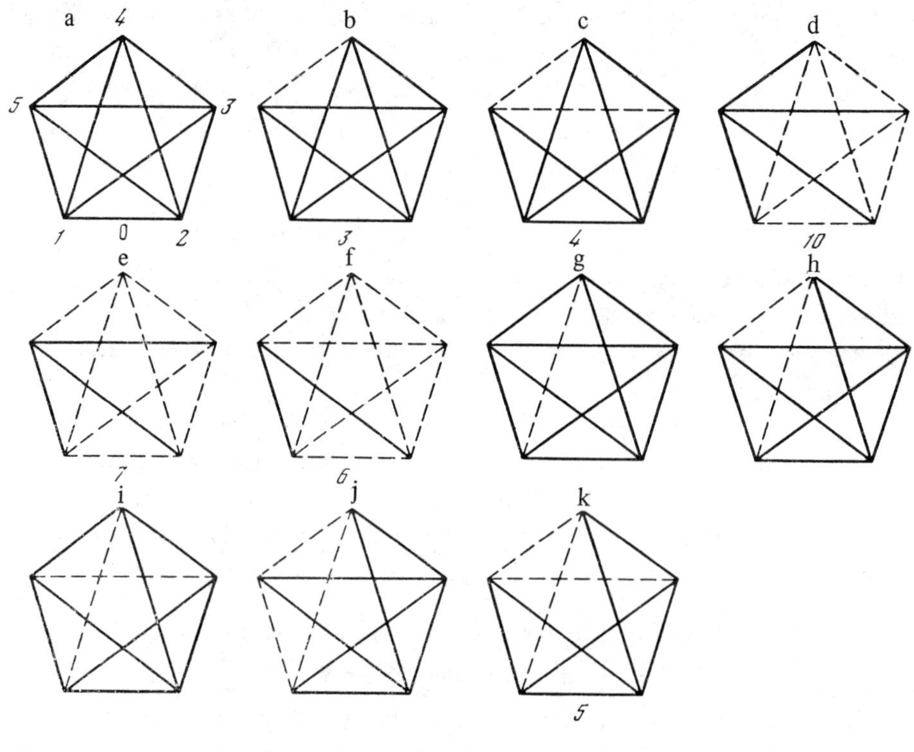

FIGURE 6

The mixed tetrahedrons may be taken to be composed of five black edges and one red edge. From a mixed tetrahedron 5 completely inequivalent diagrams are found (Figure 6g, h, i, j, k). However the first two of these are equivalent to Figure 6b, c, and the last two, after recoloring the vertices 3, 5 and then the vertex 3, are shown to be equivalent to Figure 6e, f.

In such a manner we find that there are not more than 7 mutually inequivalent 0-composite C-types. Counting the number of red triangles in each of the 7 diagrams 6a, b, c, d, k, we find that each of these diagrams is distinct (see Figure 6). Moreover, the mutual inequivalence of all of these diagrams is thereby established.

We will now show that the seven domains which we have found contain interior points. To verify this it is sufficient to put the parameters ρ_{ij} and $\rho_{\bar{i}\bar{j}}$ appearing as the weights of the pretypes of these domains (i.e. corresponding to the edges of the diagram of the given domain) equal to some number $a > 0$, and set the remaining parameters of the point equal to some $b > a$. A point having such parameters lies in the interior of the C-type domain since the inequalities defining this domain are strict.

The derivation of the 0-composite C-types is finished, and we have

PROPOSITION 6.1. *There exist exactly seven* 0*-composite C-type five-dimensional parallelohedra which are mutually inequivalent with respect to transformations of the group* $\mathfrak{G}_1$. *The diagrams corresponding to these domains are listed in Table* I, *below, as numbers* 1–7.

2. We consider the case of the 1-composite types. Each diagram corresponding to this case contains 9 edges and one b-triangle. We will number the vertices of the diagram so that the b-triangle coresponds to the vertices 4 and 5. Starting with each of the three forms of the tetrahedron 1234 as we did in subsection 1, we will add the three edges 15, 25, 35, with ends at the vertex 5. By recoloring vertex 5 these three edges may be converted either to all black or to two black and one red.

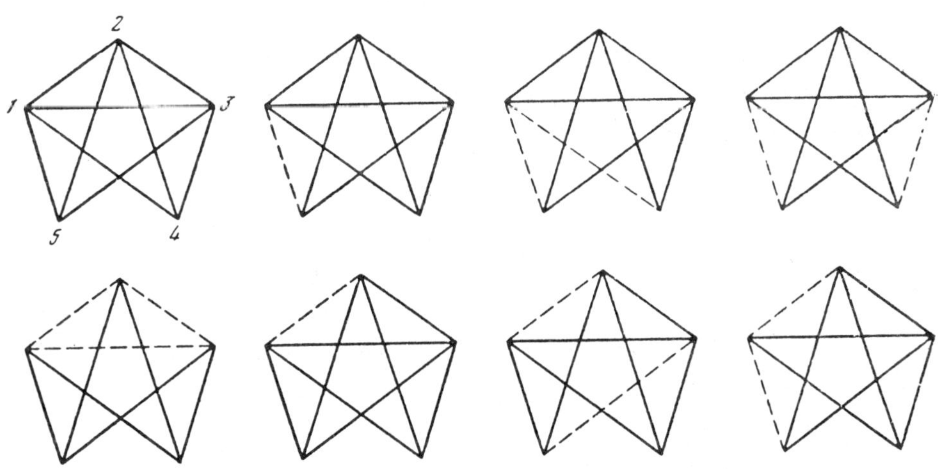

FIGURE 7

The investigation of all possible cases gives 8 inequivalent variants of a diagram having 9 edges (Figure 7). When we add to each of the variants of Figure 7 the b-triangle 123, Lemma 5.6 must be satisfied. According to Lemma 5.6, if the triangle 123 is black, then the b-triangle 123 is unique, its sides having the same color as the edges of the triangle 123. If the triangle 123 is red, then three variants of the b-triangle 123 are permissible.

Looking through all 8 of the variants of Figure 7, we see that only the last four allow three variants of b-triangles and the last three have only two mutually inequivalent variants. All in all we get 11 different 1-composite C-types, the diagrams of the domains of which are displayed in Table I, numbers 8–18. We will show that all of these domains are mutually inequivalent and have interior points.

Enumerating the number of red triangles in each of the diagrams of Figure 7, one finds that only the 3rd and 4th figures have 4 red triangles and only the 6th and 8th figures have 3 red triangles. The inequivalence of each of these two pairs of figures follows from the inequivalent arrangement of red triangles with respect to one another in the figures of each pair. In such a manner we find that diagrams constructed from different figures of Figure 7 are inequivalent. If two diagrams in Table I both arise from the same figure in Figure 7, their inequivalence is obvious.

We shall find forms (points) lying interior to the domains of each of these C-types. For this, we take the parameters ρ_{ij} and $\rho_{\overline{i}\overline{j}}$, corresponding to the edges of the given diagram, to be equal to some $a > 0$. We set the parameters corresponding to the sides of the b-triangle, which by comparison are colored differently than those of the corresponding triangle, equal to $a + \epsilon$, $\epsilon > 0$. We set the remaining parameters equal to $3(a + \epsilon)$.

PROPOSITION 6.2. *There exist exactly* 11 *domains of* 1-*composite C-types which are mutually inequivalent with respect to transformations of the group* $\mathfrak{G}_\Gamma$. *The diagrams corresponding to these domains are listed in Table* I, *numbers* 8–18.

3. From Lemma 5.4 we see that we do not lose generality in the derivation of the 2-composite types if we assume that the composite weights have pretypes of index J_{35} and J_{45}. The diagrams of such domains contain 8 edges and two b-triangles, 123 and 124.

Starting from the three inequivalent variants of the tetrahedron 1234 we find 10 mutually inequivalent eight-edged figures (Figure 8). Then, depending on the colors of the triangles 123 and 124 of these figures, we assign the b-triangles 123 and 124. The first four diagrams of Figure 8 admit a unique variant of these b-triangles; diagrams five and six admit 3 variants each, these being mutually inequivalent. The remaining 4 figures admit 5 variants each, among which there are 4 which are mutually inequivalent. All together we have 26 different diagrams of 2-composite C-types (Table I, 19–44).

It remains to be established that all of these diagrams, in fact, correspond to C-type domains. For this we shall find an interior point in each of the sets $\mu \in M_0$ corresponding to these diagrams. Excepting diagrams 31, 32, 39 and 40 of Table I, points for all the remaining 22 may be found exactly as in subsection 2 for the case of 1-composite C-types. Namely, we assume that the parameters ρ_{ij} and $\rho_{\overline{i}\overline{j}}$ corresponding to all 8 edges of the given diagram are equal to some $a > 0$. We also assume that every parameter corresponding to the side of a b-triangle colored differently from that of the corresponding triangle is equal to $a + \epsilon$, where $\epsilon > 0$, and that all of the remaining parameters are equal to $3(a + \epsilon)$. The points obtained by this method we will call *standardly assigned.*

With the exception of ρ_{35}, we define all of the parameters of the points for the C-type domains corresponding to diagrams 31, 32, 39 and 40 to be the same as those

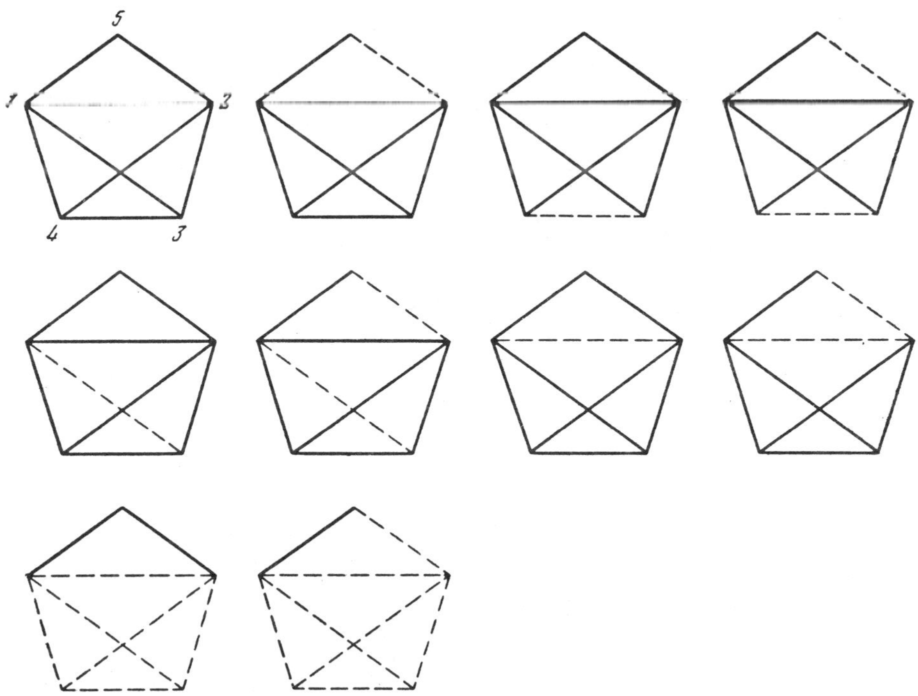

FIGURE 8

standardly assigned. We choose the parameter ρ_{35} (we have assumed that the vertices of all of the diagrams of Table I are enumerated as in the "empty" diagram) to be equal to $a + 2\epsilon$. Convincing ourselves that all points described here are really interior points of the corresponding C-type domains, we have

PROPOSITION 6.3. *There exist* 26 *and only* 26 *domains of* 2-*composite C-type parallelohedra which are mutually inequivalent with respect to the group* $\mathfrak{G}_1$. *The diagrams corresponding to these domains are listed in Table* I, *numbers* 19–44.

4. According to Corollary 2 of Lemma 5.2 (or to Lemma 5.4) it is necessary to consider two cases of 3-composite types. We begin with the case of Figure 5c in which the composite weights correspond to pretypes of index J_{ij}, J_{ik} and J_{im}. The seven edges of the diagram of such a C-type form a tetrahedron with an additional edge ip which always may be chosen black. We get all four mutually inequivalent arrangements of these seven edges in the diagrams (Figure 9, the top half).

For a more convenient enumeration of the different possible variants of arrangements of b-triangles with respect to each of the figures of the top row of Figure 9 it is better to represent the tetrahedron as it is given in the bottom row of Figure 9.

Considering different variants of arrangements of b-triangles with respect to the tetrahedron $jkmp$ we get in case a) of Figure 9 one, in case b) seven, in case c) six and in case d) two inequivalent variants. One of the seven variants of case b) is forbidden

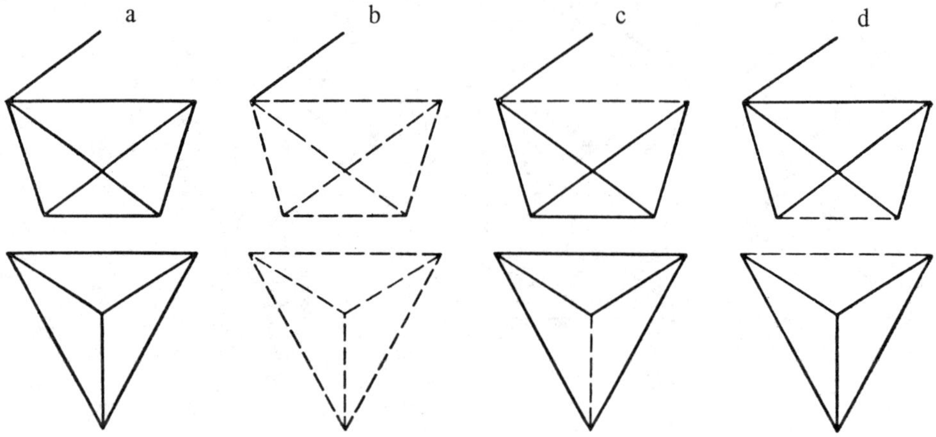

FIGURE 9

by Lemma 5.8; therefore we get altogether 15 mutually inequivalent diagrams of C-type domains. In Table I these diagrams are 45–59.

For diagrams 45–48, 51–54 and 57 we take as an example of an interior point of the corresponding C-type domain the standardly assigned point. For diagrams 49, 50 and 59 we take the standardly assigned point and alter the parameter ρ_{35}, assigning it the value $a + 2\epsilon$. For each of the C-type domains corresponding to diagrams 55, 56 and 58 examples of interior points are given by considering the following deviations from the standardly assigned: 1) for diagram 55, take $\rho_{\overline{35}} = a + 2\epsilon$; 2) for diagram 56, take $\rho_{13} = \rho_{35} = a + 2\epsilon$; 3) for diagram 59, take $\rho_{\overline{15}} = a + 1.5\epsilon$ and $\rho_{35} = a + 2\epsilon$.

In the derivations of the 3-composite types of the form 5d we assume that composite weights belong to pretypes of index J_{ij}, J_{ik}, J_{jk}. Then the triangles *kmp*, *jmp* and *imp* formed by the edges of the diagram give four inequivalent possibilities, characterized by the presence among these triangles of respectively 3, 2, 1 and 0 black triangles.

In the first case only one variant of the diagram is possible, since all of its three b-triangles can have only black sides. In the second there are two, in the third there are four, and in the fourth there are six mutually inequivalent diagrams. In this way we obtain the 13 diagrams 60–72 of Table I. Examples of interior points for these C-type domains except for 68–70 are given by the standardly assigned points. In the case of the diagrams 68–70 the standardly assigned points are changed so that the parameter ρ_{35} assumes the value $a + 2\epsilon$.

Summarizing the results obtained, we have

PROPOSITION 6.4. *There exist* 28 *and only* 28 *domains of* 3-*composite C-type which are mutually inequivalent under the group* $\mathfrak{G}_1$. *The diagrams* 45–72 *correspond to these domains.*

5. As follows from Corollary 1 of Lemma 5.2, the diagrams of the domains of the 4-composite type consist of tetrahedra and b-triangles "inscribed" in them.

If the tetrahedron is black, then all of the variants obtained by inscribing b-triangles are equivalent. In the case of red or mixed tetrahedra there are four mutually inequivalent variants. For this investigation it is convenient to take a black tetrahedron with all black edges, a mixed one with one red edge, the remaining black, and a red tetrahedron with two red edges, the remaining black. In this last case the sequence of variants is rejected on the basis of Lemma 5.8.

The 9 different diagrams of 4-composite C-types obtained are represented in Table I in diagrams 73–75 and a–f.

Examples of interior points for the C-type domains corresponding to diagrams 73–75 and a–c are given by the standardly assigned points. For d, e and f, examples are defined by the following deviations from the standardly assigned points: 1) for diagram d we assume $\rho_{15} = \rho_{25} = a + 2\epsilon$; 2) for e we assume $\rho_{12} = \rho_{\overline{35}} = a + 2\epsilon$; 3) for f we assume $\rho_{25} = a + 1.5\epsilon$ and $\rho_{\overline{35}} = a + 2\epsilon$.

PROPOSITION 6.5. *There exist* 9 *and only* 9 *domains of* 4-*composite C-types which are mutually inequivalent under the group* $\mathfrak{G}_1$. *These domains correspond to diagrams* 73–75 *and* a–f *of Table* I.

From Propositions 6.1–6.5 follows

PROPOSITION 6.6. *The perfect domain* R_1 *intersects with* 81 *C-type domains which are mutually inequivalent under* $\mathfrak{G}_1$, *i.e. there exist no more than* 81 *general primitive five-dimensional parallelohedra whose domains intersect the interior of* R_1. *The diagrams corresponding to these domains are listed in Table* I.

§7. The C-types of the domain R_2

1. The third perfect domain R_2 of the cone K for $n = 5$ is defined ([1], §44) as the set of forms

$$\begin{aligned} f(x_1, \ldots, x_5) = \rho_1 x_1^2 + \ldots + \rho_5 x_5^2 + \rho_6 (x_1 - x_3)^2 + \ldots + \rho_{11} (x_2 - x_5)^2 \\ + \rho_{12} (x_1 + x_2 - x_3 - x_4)^2 + \rho_{13} (x_1 + x_2 - x_3 - x_5)^2 + \rho_{14} (x_1 + x_2 \\ - x_4 - x_5)^2 + \rho_{15} (x_1 + x_2 - x_3 - x_4 - x_5)^2, \end{aligned} \tag{7.1}$$

where $\rho_1, \ldots, \rho_{15} \geqslant 0$.

Taking advantage of the linear transformation

$$\begin{aligned} &x_1 \to x_1 + x_3, && x_2 \to x_2 + x_3, \quad x_3 \to -x_4 - x_5, \\ &x_4 \to x_1 + x_2 + x_3 + x_4, && x_5 \to x_1 + x_2 + x_3 + x_5, \end{aligned} \tag{7.2}$$

the forms of R_2 may be written as

$$f(x_1, \ldots, x_5) = \sum_{\substack{p, q = 0 \\ (p < q)}}^{5} \sigma_{pq} (x_p + x_q)^2, \qquad x_0 = -\sum_{i=1}^{5} x_i, \quad \sigma_{pq} \geqslant 0. \tag{7.3}$$

TABLE I

C-types of the domain R_1 (diagrams)

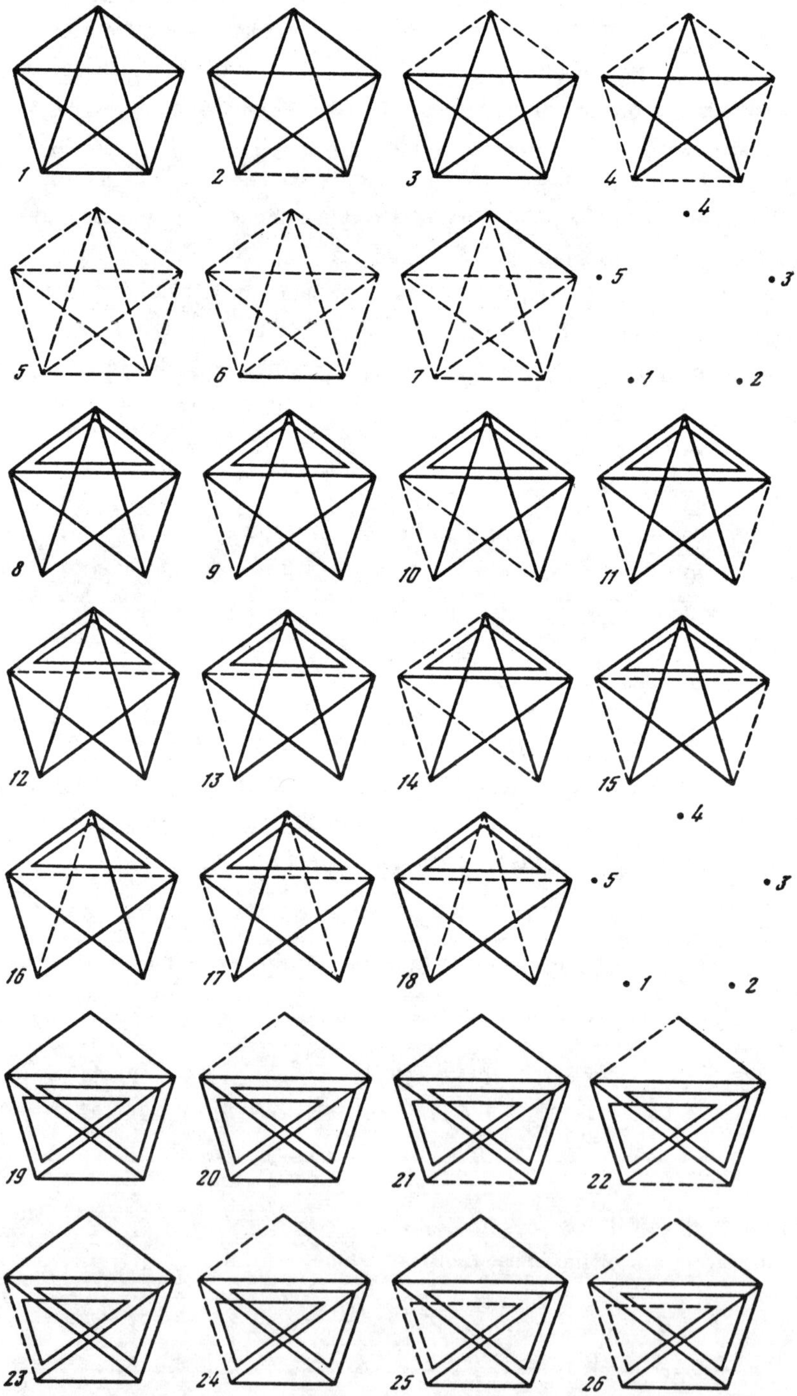

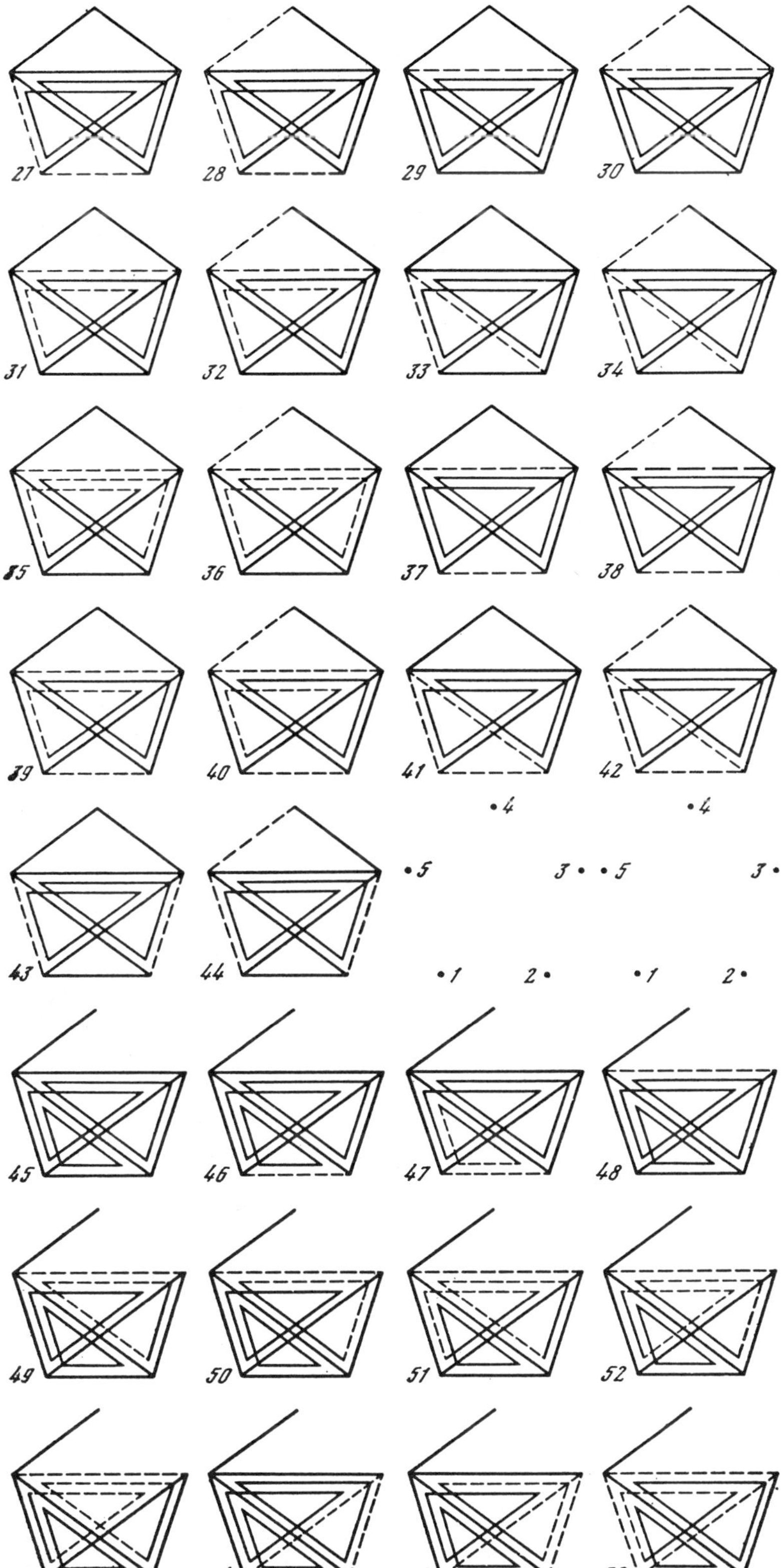

27
28
29
30
31
32
33
34
35
36
37
38
39
40
41
42
43
44
45
46
47
48
49
50
51
52
53
54
55
56
4 5 3 1 2
4 5 3 1 2

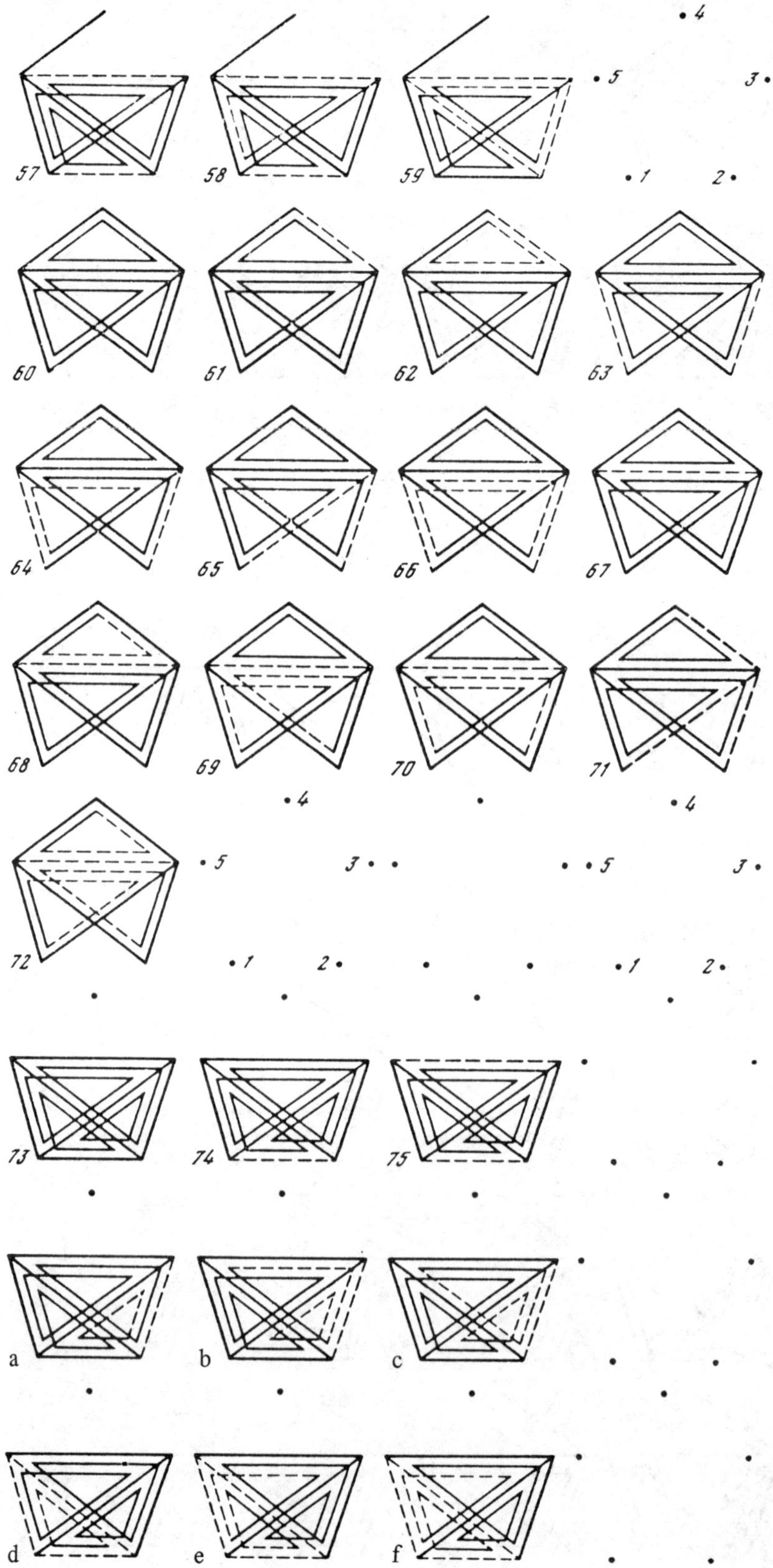
4
5
3
1
2
57
58
59
60
61
62
63
64
65
66
67
68
69
70
71
72
73
74
75
a
b
c
d
e
f

Here and everywhere below, if the contrary is not stipulated, when we refer to R, we will mean the transformed domain, i.e. the set of forms (7.3). The parameters $\sigma_{pq} = \sigma_{qp}$ ($p, q = 0, 1, \ldots, 5; p \neq q$) are related to the parameters $\rho_1, \ldots, \rho_{15}$ of the form (7.1) by the equalities

$$\sigma_{01} = \rho_9, \quad \sigma_{02} = \rho_6, \quad \sigma_{03} = \rho_{14}, \quad \sigma_{04} = \rho_5, \quad \sigma_{05} = \rho_4,$$
$$\sigma_{12} = \rho_{15}, \quad \sigma_{13} = \rho_1, \quad \sigma_{14} = \rho_{10}, \quad \sigma_{15} = \rho_{11}, \quad \sigma_{23} = \rho_2,$$
$$\sigma_{24} = \rho_7, \quad \sigma_{25} = \rho_8, \quad \sigma_{34} = \rho_{13}, \quad \sigma_{35} = \rho_{12}, \quad \sigma_{45} = \rho_3.$$

Let $\{\Gamma\}$ be the set of lattices corresponding to the forms (7.1) of R_2. Since the determinant of the transformation is equal to 2, as in §3 the coordinates of a lattice point in a frame determined by one of the forms (7.3) are either of the form $(l_1, \ldots, l_5)$ or of the form $(l_1 + ½, \ldots, l_5 + ½)$, where $l_1, \ldots, l_5$ are arbitrary integers. The transformation (7.2) transforms the set of indices I_σ ($\sigma = 1, \ldots, 31$) into the set of indices J_σ (see (3.10)–(3.12)).

We will now show that R_2, considered as the set of forms (7.3), transforms into itself under an arbitrary permutation of the indices 0, 1, . . . , 5 of the parameters of these forms. This is obvious if 0 is left fixed. Consider the unimodular transformation

$$\begin{aligned} &x_i \to -x_1 - x_2 - x_3 - x_4 - x_5 = x_0, \\ &x_j \to x_j, \qquad j = 1, \ldots, 5, \quad j \neq i, \end{aligned} \tag{7.4}$$

where i has one of the values 1, . . . , 5. Note that under this transformation $x_0 = -\Sigma_{j=1}^5 x_j \longrightarrow x_i$. Thus the forms of the type (7.3) are left invariant under this transformation. The transformation (7.4) corresponds to a permutation of the indices 0 and i. We remark that these transformations generate the entire group $\mathfrak{G}_2$ of automorphisms of R_2.

According to [1], the domain R_2 is bounded by fifteen 14-dimensional faces. Each of these faces is defined by one of the equations $\sigma_{pq} = 0$, $p, q = 0, 1, \ldots, 5$, $p < q$.

Denote by l^i ($i = 1, \ldots, 5$) a vector for which $l_i = 1$ and the remaining coordinates are zero; by $l^{\overline{ij}}$ ($i, j = 1, \ldots, 5; i < j$) a vector for which $l_i = -l_j = 1$ and the remaining coordinates are zero; and by d^{ij} ($i, j = 1, \ldots, 5; i < j$) a fractional vector where $l_i = l_j = -½$ and the remaining coordinates are ½.

None of the 25 vectors thus enumerated are congruent mod 2, i.e., they define 25 indices. For the remaining six indices we adopt the notation $J_0 = (½, ½, ½, ½, ½)$, $J_1 = (-½, ½, ½, ½, ½, ½), \ldots, J_5 = (½, ½, ½, ½, -½)$.

We will investigate the pretypes of forms $f \in R_2$ belonging to each of the 31 five-dimensional indices.

THEOREM 7.1. *Corresponding to each index, with the exception of the six indices $J_0, J_1, \ldots, J_5$, the interior points of R_2 and the interior points of its 14-dimensional faces belong to a single pretype. These pretypes are given by l^i, $l^{\overline{ij}}$ and d^{ij} $(i, j = 1, \ldots, 5; i < j)$.*

PROOF. Since R_2 transforms into itself under an arbitrary permutation of the indices 0, 1, . . . , 5 of the parameters σ_{pq}, without loss of generality we can carry through the proof for the three indices $J' = (1, 0, 0, 0, 0)$, $J'' = (1, 1, 0, 0, 0)$ and $J''' = (-½, -½, ½, ½, ½)$.

The pretypes of index J' have the form $l' = (1 + l_1, l_2, \ldots, l_5)$, where $l_1, \ldots, l_5$ are either all even or all odd whole numbers. Substituting l' into (7.3), we have

$$f(l') = \sum_{r=2}^{5} [\sigma_{0r}(1 + l_1 + \ldots + l_5 - l_r)^2 + \sigma_{1r}(1 + l_1 + l_r)^2] \\ + \sigma_{01}(l_2 + l_3 + l_4 + l_5)^2 + \sum_{\substack{r,s=2 \\ (r<s)}}^{5} \sigma_{rs}(l_r + l_s)^2. \tag{7.5}$$

Since the numbers $l_1, \ldots, l_5$ are congruent mod 2, the following inequality quickly follows from (7.5):

$$f(l') \geqslant \sum_{r=2}^{5} (\sigma_{0r} + \sigma_{1r}) = f(l^1). \tag{7.6}$$

At interior points of R_2 we have $\sigma_{pq} > 0$ $(p, q = 0, 1, \ldots, 5; p < q)$; at interior points of one of its 14-dimensional faces only one of these parameters is equal to zero, the others being strictly positive. On the basis of this, it follows from (7.5) that at interior points of R_2 and at interior points of its 14-dimensional faces, (7.6) becomes an equality if and only if $l_1 = l_2 = \cdots = l_5 = 0$. This establishes the theorem for the index J'.

The pretypes of index J'' have the form $l'' = (1 + l_1, -1 + l_2, l_3, l_4, l_5)$, where $l_1, \ldots, l_5$ are congruent mod 2. The substitution of l'' into (7.3) gives

$$f(l'') = \sigma_{01}(1 + l_0 + l_1)^2 + \sigma_{02}(-1 + l_0 + l_2)^2 + \sum_{t=3}^{5} [\sigma_{1t}(1 + l_1 + l_t)^2 \\ + \sigma_{2t}(-1 + l_2 + l_t)^2] + \sigma_{12}(l_1 + l_2)^2 + \sum_{\substack{t,u=3 \\ (t<u)}}^{5} \sigma_{tu}(l_t + l_u)^2 + \sum_{t=3}^{5} \sigma_{0t}(l_0 + l_t)^2, \tag{7.7}$$

where $l_0 = -\Sigma_{i=1}^{5} l_i$. From (7.7) we find that

$$f(l'') \geqslant \sigma_{01} + \sigma_{02} + \sum_{t=3}^{5} (\sigma_{1t} + \sigma_{2t}) = f(l^{\overline{12}}). \tag{7.8}$$

As in the case of the pretypes l', we deduce from the above arguments that for all pretypes l'', with the exception of the pretype $l'' = l^{\overline{12}}$, inequality (7.8) is strict for all these points of R_2 considered in our theorem.

The pretypes of index J''' have the form $l''' = (l_1 - ½, l_2 - ½, l_3 + ½, l_4 + ½, l_5 + ½)$, where $l_1, \ldots, l_5$ are whole numbers congruent mod 2. After the substitution of l''' into (7.3) we have

$$f(l''') = \sum_{\substack{g,\,h=0\\(g<h)}}^{2} \sigma_{gh}(l_g + l_h - 1)^2 + \sum_{g=0}^{2}\sum_{t=3}^{5} \sigma_{gt}(l_g + l_t)^2$$
$$+ \sum_{\substack{t,\,u=3\\(t<u)}}^{5} \sigma_{tu}(l_t + l_u + 1)^2 \geqslant \sigma_{01} + \sigma_{02} + \sigma_{12} + \sigma_{34} + \sigma_{35} + \sigma_{45} = f(d^{12}). \tag{7.9}$$

From (7.9) we can see that in the case of forms f where not more than one of the coefficients σ_{pq} is equal to zero the equality $f(l''') = f(d^{12})$ is achieved if and only if $l''' = d^{12}$. This means that all the interior points of R_2 and of its 14-dimensional faces belong to the pretype d^{12} of index J'''. With this, the proof of Theorem 7.1 is finished.

We now consider pretypes for forms in R_2 which correspond to each of the indices $J_0, J_1, \ldots, J_5$.

First of all we remark that the collection of indices $J_0, J_1, \ldots, J_5$ is invariant and transitive with respect to renumbering the coordinates and the transformations (7.4), where these latter transformations induce automorphisms of R_2 in E^N. Therefore it is sufficient to carry through the investigation for one of the given indices, for example, J_1.

Pretypes of index J_1 have the form $d' = (l_1 - \frac{1}{2}, l_2 + \frac{1}{2}, \ldots, l_5 + \frac{1}{2})$, where $l_1, \ldots, l_5$ are whole numbers congruent mod 2. The value of the form (7.3) for d^1 is

$$f(d^1) = \sigma_{01}(l_0 + l_1 - 2)^2 + \sum_{r=2}^{5} [\sigma_{0r}(l_0 + l_r - 1)^2 + \sigma_{1r}(l_1 + l_r)^2]$$
$$+ \sum_{\substack{r,\,s=2\\(r<s)}}^{5} \sigma_{rs}(l_r + l_s + 1)^2, \tag{7.10}$$

where $l_0 = -\Sigma_{i=1}^{5} l_i$.

The fact that $l_0, \ldots, l_5$ are congruent mod 2 implies the inequalities

$$|l_0 + l_r - 1| \geqslant 1, \quad |l_r + l_s + 1| \geqslant 1, \quad |l_1 + l_r| \geqslant 0, \quad |l_0 + l_1 - 2| \geqslant 0$$
$$(r,\ s = 2, \ldots, 5;\ r \neq s). \tag{7.11}$$

For admissible values of $l_1, \ldots, l_5$ not all of the inequalities of (7.11) may simultaneously become equalities, since the left-hand side of at least one of the last five inequalities must equal 2. The outcome of this is that we get the following five inequalities for the form $f(d^1)$ from (7.10):

$$f(d_1) \geqslant 4\xi + \sum_{r=2}^{5} \sigma_{0r} + \sum_{\substack{r,\,s=2\\(r<s)}}^{5} \sigma_{rs}, \tag{7.12}$$

where $\xi = \sigma_{01}, \sigma_{12}, \sigma_{13}, \sigma_{14}, \sigma_{15}$.

It is easy to establish that of the five inequalities (7.12) only one, corresponding to a single pretype, becomes an equality at interior points of R_2 and at interior points of any of its 14-dimensional faces. Namely, the value $\xi = \sigma_{1q}$ $(q = 0, 2, 3, 4, 5)$ corresponds to the pretype defined by the numbers $l_q = -2, l_r = 0$ $(r = 0, 1, \ldots, 5; r \neq q)$.

Thus the form $f \in R_2$ of index J_1 belongs, depending on the interrelationships of the parameters $\sigma_{01}, \sigma_{12}, \sigma_{13}, \sigma_{14}, \sigma_{15}$, to one or several of the following pretypes:

$$J_1, \left(-\frac{1}{2}, -\frac{3}{2}, \frac{1}{2}, \frac{1}{2}, \frac{1}{2}\right), \left(-\frac{1}{2}, \frac{1}{2}, -\frac{3}{2}, \frac{1}{2}, \frac{1}{2}\right),$$
$$\left(-\frac{1}{2}, \frac{1}{2}, \frac{1}{2}, -\frac{3}{2}, \frac{1}{2}\right), \left(-\frac{1}{2}, \frac{1}{2}, \frac{1}{2}, \frac{1}{2}, -\frac{3}{2}\right).$$

Our proof for the index J_1 may be extended to that for the remaining indices $J_0, J_2, \ldots, J_5$.

We denote by d_q^p $(p, q = 0, 1, \ldots, 5; p \neq q)$ the coordinates of fractional vectors where $l_p = -½, l_q = -3/2, l_0 = -\Sigma_{i=1}^5 l_i$ and the remaining coordinates equal ½. We remark that $d_0^p = J_p$, $p = 0, 1, \ldots, 5$.

We have the following lemma.

LEMMA 7.2. *For each of the indices J_p $(p = 0, 1, \ldots, 5)$ the forms $f \in R_2$ belong to one or several of the five pretypes d_q^p $(q = 0, 1, \ldots, 5; q \neq p)$ depending on which of the five parameters σ_{pq} $(q = 0, 1, \ldots, 5; q \neq p)$ of the form f is the smallest.*

If the form f belongs to the interior of R_2 or is an interior point of any of the 14-dimensional faces of R_2, then it may not belong to any pretype of index J_p other than one of the five listed above.

2. From Theorem 7.1 and Lemma 7.2 it follows, first, that R_2 intersects with no more than five C-type domains and, second, that the 14-dimensional faces of R_2 intersect with faces of C-type domains in sets of dimension not greater than 13. The latter means that all of the C-type domains of the perfect domain R_2 have common interior points with those of the perfect domains of the cone $\boldsymbol{K}$ which are contiguous with R_2 at a face of dimension 14. It is well known [1], [28] that all of the perfect domains which are contiguous with R_2 are equivalent to R_1. Thus we have the following theorem.

THEOREM 7.3. *All general C-types of five-dimensional parallelohedra, except for the C-type of the domain R (the "first type"), may be placed in correspondence with arithmetical C-types which have interior points in R_1.*

COROLLARY. *Each C-type domain, with the exception of the first type, is represented by one or several diagrams (symbolic strings) each corresponding to its intersection with R_1.*

Thus the perfect domain R_2 does not yield any new general C-types in addition to those found for R and R_1. However, in order to answer the question of the mutual inequivalence of the C-type domains intersecting R_1, and also for the determination of the L-types of R_2, it is necessary to consider the subdivision of R_2 into C-type domains.

3. As has been discussed in subsection 1, the group of automorphisms of R_2 given by the forms (7.3) is the group of permutations of the indices of the parameters. From this it follows that two C-type domains intersecting R_2 will be equivalent under the group $\mathfrak{G}_2$ if and only if one is transformed into the other under such a permutation.

Corresponding to each C-type domain of R_2 we introduce the following geometrical diagrams. We number the vertices of a regular hexagon with the numbers 0, 1, . . . ,5. Each of the 6 indices J_p ($p = 0, 1, \ldots, 5$) of the given C-type domain corresponds to the parameters $\sigma_{pr} < \sigma_{ps}$, $s = 0, 1, \ldots, 5$, $s \neq p \neq r$, of the form $f \in R_2$. We represent the parameter σ_{pr} in our diagram by a segment joining the vertices p and r which has an arrow pointing towards the vertex p. Thus the diagrams of these C-type domains consist of the vertices of a regular hexagon and six arrows directed to these vertices (with one and only one arrow directed to each vertex). In addition there are segments bearing these arrows. Each of these segments bears one or two arrows, and therefore the number of segments in a diagram may vary between 3 and 6 (see Table II).

We will designate by τ ($\tau = 1, 2, \ldots, 21$) both the intersection of the C-type domains with R_2 and the diagram of Table II to which it corresponds.

We will decipher, for example, diagram 1 of Table II. The corresponding C-type domain, i.e. domain 1, is determined by the inequalities

$$\sigma_{01} < \sigma_{02}, \sigma_{03}, \sigma_{04}, \sigma_{05}; \qquad \sigma_{01} < \sigma_{12}, \sigma_{13}, \sigma_{14}, \sigma_{15};$$
$$\sigma_{23} < \sigma_{02}, \sigma_{12}, \sigma_{24}, \sigma_{25}; \qquad \sigma_{23} < \sigma_{03}, \sigma_{13}, \sigma_{34}, \sigma_{35};$$
$$\sigma_{45} < \sigma_{04}, \sigma_{14}, \sigma_{24}, \sigma_{34}; \qquad \sigma_{45} < \sigma_{05}, \sigma_{15}, \sigma_{25}, \sigma_{35}.$$

We will now verify that there are no C-type domains corresponding to diagrams which contain closed polygons. Let us assume the contrary, that a diagram contains a closed polygon, consisting of the segments $p_1p_2, p_2p_3, \ldots, p_rp_1$ ($p_1, \ldots, p_r$ are the vertices of this polygon). Then, depending on the placement of the arrows (there is only a single arrow directed towards each vertex), either the inequalities $\sigma_{p_1p_2} < \sigma_{p_2p_3} < \cdots < \sigma_{p_rp_1} < \sigma_{p_1p_2}$ or the contrary ones must hold. The contradiction contained in either of these systems of inequalities proves the assertion.

For the determination of the C-type domains which are mutually inequivalent with respect to transformations of the group $\mathfrak{G}_2$ it is sufficient to find all of those diagrams (not containing closed polygons) which are inequivalent with respect to a permutation of the vertices. Elementary considerations yield the 21 domains whose diagrams are represented in Table II.

It is easy to verify that each of the domains defined by the diagrams of Table II

TABLE II

Diagrams of the C-types of the domain R_2

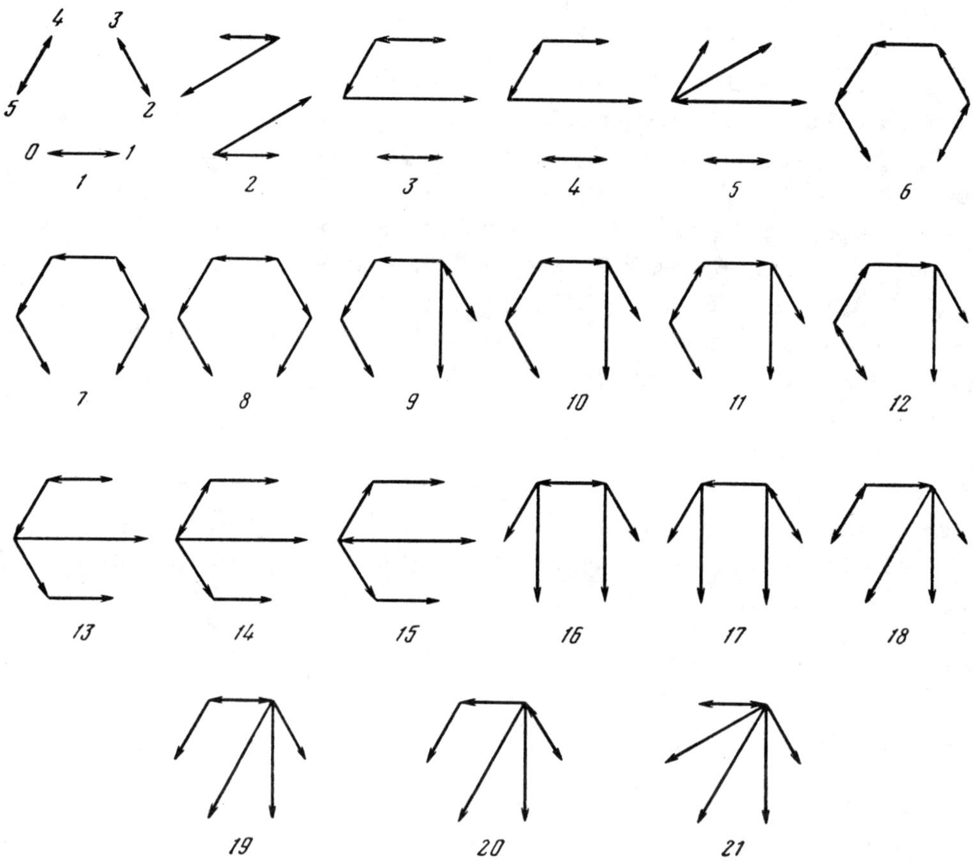

are in fact domains, i.e. they have interior points. To do this, we shall display an interior point for each domain. We shall call a segment of the diagram *primary* if it has an arrow on both ends. A diagram may have from 1 to 3 such segments. We will call a segment whose end not having an arrow is coincident with an end of a primary segment a *secondary* segment; and, continuing, we have in general a *k-segment.*

For each of the diagrams of Table II the element having the following values for its parameters σ_{pq} is interior to the corresponding C-type domain: parameters corresponding to k-segments are taken equal to k; the remaining parameters which are not represented by segments with arrows are chosen to be equal to 6 (the number 6 is chosen since it is greater than 5 and 5-segments do not occur).

We now formulate the results we have obtained.

PROPOSITION 7.4. *Among the C-type domains intersecting the domain R_2 there are* 21 *and only* 21 *which are mutually inequivalent under the group $\mathfrak{G}_2$ of automorphisms of this domain.*

PROPOSITION 7.5. *There exist no more than* 21 *general C-types whose domains intersect with the domain* R_2; *the diagrams of the intersections of these C-types with* R_2 *are displayed in Table* II.

§8. The *C*-types of the five-dimensional parallelohedra: Concluding list

1. In §§6 and 7, we found all of the *C*-type domains which intersect the domains R_1 and R_2. We selected from these subsets which were mutually inequivalent with respect to the groups $\mathfrak{G}_1$ and $\mathfrak{G}_2$ respectively. Here we will investigate the question of equivalence of these *C*-type domains with respect to the entire group $\mathfrak{G}$. Thus, we will finally arrive at the realization of the four-part algorithm for deducing the *C*-types (§1).

Let $R_i \subset K_n$ be some perfect domain and let $\mathfrak{G}_i \subset \mathfrak{G}$ be the group of its automorphisms. Let $\delta_1 = \Delta_1 \cap R_i$ and $\delta_2 = \Delta_2 \cap R_i$ be intersections of the domain R_i with two *C*-type domains Δ_1 and Δ_2 which are inequivalent with respect to the group $\mathfrak{G}_i$. There are two possibilities:

1) The sets $\Delta_1 \backslash \delta_1$ and $\Delta_2 \backslash \delta_2$ do not have interior points in the same or in mutually equivalent perfect domains of the cone K_n.

2) They do have such points.

It is obvious that in the first case the *C*-types of the parallelohedra corresponding to Δ_1 and Δ_2 are different.

If the intersections of $\Delta_1 \backslash \delta_1$ and $\Delta_2 \backslash \delta_2$ (or the intersections of sets equivalent to them) with some perfect domain R_j are equivalent under the group of automorphisms of this domain, then the domains Δ_1 and Δ_2 are equivalent, i.e., only a single *C*-type parallelohedron will correspond to them.

Excepting this, if the domain R_j is equivalent to R_i, then the *C*-type domains Δ_1 and Δ_2 are equivalent if for a mapping of R_i into R_j the domain δ_1 is mapped into $\Delta_2 \cap R_j$ and δ_2 into $\Delta_1 \cap R_j$.

2. We pass to the investigation of the 5-dimensional case. From subsection 1, from the connectedness of the *C*-type domains and from Theorem 7.3 it follows that for the investigation of equivalence with respect to the group $\mathfrak{G}$ we need to consider only those *C*-type domains intersecting R_1 which are not wholly contained in it and which have interior points on one of its 14-dimensional faces.

The domain R_1 is contiguous at 14-dimensional faces to domains which are equivalent to the domains R, R_2 and to itself [1], [28].

Since R is comprised of an entire *C*-type domain (§3), faces at which R_1 abuts with domains equivalent to R are "not penetrated", i.e. they consist only of boundary points of *C*-type domains.

The domain R_1 is contiguous with domains which are equivalent to it at faces of two types. A representatve of the first type is, for example, the face F, given (see (3.8)) by the equation

$$\rho_{12}+\rho_{13}+\rho_{14}+\rho_{15}=0. \tag{8.1}$$

According to Theorem 3.2, if a form $f \in F$ belongs to two or more pretypes of index J_q^d, then it is not in general position. It follows that the face F and faces equivalent to it consist of boundary points of C-type domains, and the C-type domains intersecting these faces do not penetrate into the perfect domains lying on either side.

3. The second type of face at which R_1 is contiguous with domains equivalent to it comprises faces equivalent to the face D:

$$\rho_{12}+\rho_{13}+\rho_{14}+\rho_{23}+\rho_{24}+\rho_{\overline{34}}=0. \tag{8.2}$$

Since the forms of R_1 are written as nonnegative decompositions as in (3.4), all of the parameters in (8.2) are equal to zero on the face D. Since the parameters equal to zero are the smallest by comparison with the rest, the diagrams corresponding to C-type domains belonging to the face D and those equivalent to it contain a mixed tetrahedron with two b-triangles inscribed in its black triangles.

By considering the set of diagrams of Table I, we find that the following eight C-type domains occur at the face D:

$$\begin{array}{llll} aaaaaaab00, & aaaaaabb00, & aaaaaa0b00, & aaaaaa2b00, \\ aaa0aa0b00, & aaa3aa0b00, & aaa3aa3b00, & aaa3aa2b00. \end{array} \tag{8.3}$$

These correspond to diagrams 21, 22, 46, 47, 74, a, b and c.

We will now show that each of the domains (8.3) has interior points on D. First of all, from Theorems 3.2 and 3.3 we know that all of the points of D, except those points where D intersects faces either of the form F_ν or H_ν, are in general position with respect to the pretypes belonging to the 21 indices mentioned in these theorems. Further, according to Lemma 5.1, of the remaining ten indices only pretypes of the list (5.1) contain interior points on D. The latter easily follows by considering the symbolic strings for each of the domains (8.3) and then the corresponding inequalities defining these domains.

We will investigate the mutual equivalence of the eight C-type domains whose intersections with R_1 are listed in (8.3). For this we transform the domain R_1' contiguous with R_1 at D into the domain R_1 using the transformation

$$x_1 \to x_4, \quad x_2 \to x_5, \quad x_3 \to -x_1-x_2+x_3+x_4+x_5, \quad x_4 \to x_1, \quad x_5 \to x_2. \tag{8.4}$$

After this transformation the intersections of the C-type domains (8.3) with R_1' translate respectively into the following domains:

$$\begin{array}{llll} aaa3aa2b00, & aaa3aa3b00, & aaa3aa0b00, & aaaaaa2b00, \\ aaa0aa0b00, & aaaaaa0b00, & aaaaaabb00, & aaaaaaab00. \end{array} \tag{8.5}$$

First of all, we note that the list (8.5) contains the same symbolic strings as (8.3), i.e. the transformation (8.4) acts so that the 8 intersections of R_1 with the C-type domains listed in (8.3) are left invariant. By comparing corresponding pairs of sets of the lists (8.3) and (8.5) we conclude that not more than five of the domains listed in (8.3) are mutually inequivalent.

In fact, each of three pairs of domains

$$\begin{array}{ll} aaaaaaab00, & aaa3aa2b00, \\ aaaaaabb00, & aaa3aa3b00, \\ aaaaaa0b00, & aaa3aa0b00, \end{array} \tag{8.6}$$

correspond to intersections of R_1 with two C-type domains which are equivalent with respect to the transformation (8.4); that is, those three pairs yield only three distinct C-types.

The intersections $aaaaaa2b00$ and $aaa0aa0b00$ are invariant under the transformation (8.4).

Since the five C-type domains we have isolated are inequivalent with respect to the group $\mathfrak{G}_1$ and since the transformation (8.4) leaves the union of the perfect domains R_1 and R'_1 invariant, these domains are generally mutually inequivalent with respect to the group $\mathfrak{G}$. We remark that we have here a complete description of these five C-type domains in the perfect subdivision.

From the equivalence of the pairs of sets (8.6) follows

PROPOSITION 8.1. *The domains of* 4*-composite C-types corresponding to diagrams* a, b *and* c *of Table* I *do not define new C-types which are distinct from the first* 75 *C-types enumerated in Table* I. *Namely, the C-types with diagrams* a, b *and* c *coincide respectively with the C-types corresponding to diagrams* 46, 22 *and* 21.

4. We turn now to the consideration of the C-type domains which intersect either the perfect domains which are equivalent to R_1 or the perfect domains which are equivalent to R_2.

One of the 14-dimensional faces of R_1 which adjoins a domain equivalent to R_2 is given by the equation

$$\rho_{13}+\rho_{14}+\rho_{15}+\rho_{\overline{34}}+\rho_{\overline{35}}+\rho_{\overline{45}}=0. \tag{8.7}$$

We denote this face by Q. Namely, R_1 adjoins R_2 at the face Q. The remaining faces at which R_1 adjoins domains equivalent to R_2 are equivalent to Q under the group $\mathfrak{G}_1$.

Since the forms of R_1 are written as nonnegative decompositions using (3.4), all of the parameters in (8.7) are zero on the face Q. Since the parameters equal to zero are the smallest, the diagrams of C-types appearing at the face Q contain the tetrahedron 1345 with edges 12, 14, 15 black and edges 34, 35, 45 red. This tetrahedron is red. From this it follows that C-type domains intersecting both R_1 and R_2, and only those, contain a red tetrahedron in their diagram.

Looking at Table I, we see that red tetrahedrons are contained in the diagrams of the following 24 intersections of C-type domains with R_1 (we cite the symbolic strings in such a way that the red tetrahedrons of their diagrams correspond to the face Q):

$aaaabbbbbb$	5	21	$aaaaa30bbb$	43	8
$aaaaaaabbb$	6	19	$aaaab30bbb$	44	15
$aaaaaabbbb$	7	16	$aaaa000bbb$	54	18
$aaaaaa0bbb$	16	10	$aaaa002bbb$	55	13
$aaaaab0bbb$	17	14	$1aaaa00bbb$	56	12
$aaaabb0bbb$	18	20	$3aaaa30bbb$	57	3
$aaaab22bbb$	37	5	$aaaa032bbb$	58	6
$aaaaa22bbb$	38	4	$2aaaa00bbb$	59	2
$aaaab20bbb$	39	9	$1aaa022bbb$	75	1
$aaaaa20bbb$	40	7	$1aaa000bbb$	d	5
$aaaaa00bbb$	41	11	$2aaa002bbb$	e	4
$aaaab00bbb$	42	17	$1aaa002bbb$	f	3

(8.8)

In the list (8.8) two numbers stand after the symbolic string. The first of these is the number of its diagram (or the number of the diagram equivalent to it) in Table I. The second number is the number of the corresponding diagram in Table II. We now turn to an explanation of this latter number.

Consider the following transformation of the variables $x_1, \ldots, x_5$ of the forms (3.4) of R_1:

$$x_1 \to \frac{x_1+x_2}{2}+x_3, \quad x_2 \to \frac{x_1-x_2}{2}, \quad x_3 \to \frac{x_1+x_2}{2}+x_3+x_4+x_5,$$
$$x_4 \to -\frac{x_1+x_2}{2}-x_4, \quad x_5 \to -\frac{x_1+x_2}{2}-x_5. \tag{8.9}$$

(8.9) is the composition of the inverse transformation of (3.3) and the transformation (7.2).

(8.9) maps the forms (3.4) lying on the face Q into the forms (7.3). Using this transformation, the correspondence between the diagrams of the intersections of the C-type domains with R_1 and the diagrams of the intersections of the C-type domains with R_2 of Table II is revealed.

This transformation, which induces a mapping of the diagrams of the C-type domains of R_1 into the diagrams of R_2, establishes the following correspondence between the coefficients ρ_{ij} and $\rho_{\overline{ij}}$ of the forms (3.4) and the coefficients σ_{pq} of the forms (7.3) for elements of Q:

$$\rho_{13} = \rho_{14} = \rho_{15} = \rho_{\overline{34}} = \rho_{\overline{35}} = \rho_{\overline{45}} = \sigma_{12} = 0,$$

$$\rho_{12} = \sigma_{13}, \quad \rho_{23} = \sigma_{02}, \quad \rho_{24} = \sigma_{24}, \quad \rho_{25} = \sigma_{25}, \quad \rho_{34} = \sigma_{35}, \quad \rho_{35} = \sigma_{34}, \quad \rho_{45} = \sigma_{03},$$

$$\rho_{\overline{12}} = \sigma_{23}, \quad \rho_{\overline{13}} = \sigma_{45}, \quad \rho_{\overline{14}} = \sigma_{05}, \quad \rho_{\overline{15}} = \sigma_{04}, \quad \rho_{\overline{23}} = \sigma_{01}, \quad \rho_{\overline{24}} = \sigma_{14}, \quad \rho_{\overline{25}} = \sigma_{15}.$$

The numbers τ ($\tau = 1, 2, \ldots, 21$) of the diagrams of Table II, standing opposite to the symbolic strings of the list (8.8), indicate the C-type domains of R_2 into which the C-type domains of R_1 defined by the given symbolic string are mapped by the transformation (8.9).

From (8.8) we see that the domains 5, 4 and 3 of R_2 correspond both to the sets *aaaab*22*bbb*, *aaaaa*22*bbb* and 3*aaaa*30*bbb* with diagrams 37, 38 and 57, and to the sets 1*aaa*000*bbb*, 2*aaa*002*bbb* and 1*aaa*002*bbb* with diagrams d, e and f of R_1. This means that each of the first three intersections is equivalent under $\mathfrak{G}_2$ to the corresponding member of the second three intersections.

Thus we have

PROPOSITION 8.2. *The* 4-*composite C-type domains corresponding to diagrams* d, e *and* f *do not define new C-types which are not represented by the diagrams* 1–75 *of Table* I. *Namely, the C-type domains whose intersections with* R_1 *have diagrams* d, e *and* f *are equivalent, respectively, to the C-type domains whose intersections have diagrams* 37, 38 *and* 57 *of Table* I.

Since its diagram contains 5 red tetrahedra, the domain *aaaabbbbbb* extends through 5 faces which are equivalent to the face Q not only under $\mathfrak{G}_1$ but also under the group of automorphisms of the domain *aaaabbbbbb* itself. From symmetry considerations we can see that each of these five extensions into R_2 or its equivalents is equivalent to domain 21.

Further, we find that the domains *aaaaaaabbb* and *aaaabb*0*bbb* each have two extensions at faces equivalent to Q. These two faces are equivalent to Q both under $\mathfrak{G}_1$ and under the group of automorphisms of the domain itself. Domain *aaaaaaabbb* extends in both cases into domains equivalent to domain 19. In the case of *aaaabb*0*bbb* both extensions are equivalent to 20.

Each of the remaining intersections of C-type domains with R_1 of the list (8.8) have only a single protrusion at Q and consequently have only a single extension into R_2.

We consider now how the intersections of the C-type domains with R_2 extend into R_1, i.e. the domains τ whose diagrams are given in Table II. It is easy to see that the number of protrusions of the given domain τ at the faces of R_2 is equal to the number of edges with arrows at both ends in its diagram. In fact, the parameters σ_{pq} corresponding to such edges are the only ones that vanish at the boundary of τ.

By considering the diagrams of Table II we can see that domain 1 alone has three extensions at faces which are equivalent to Q. These faces are equivalent to Q both under $\mathfrak{G}_2$ and under the group of automorphisms of domain 1. From symmetry considerations or directly we can see that all three extensions of domain 1 into R_1 or domains equivalent to it are equivalent to the domain $1aaa022bbb$. Since the domain $1aaa022bbb$ has only a single extension at the face Q, domain 1 and its three extensions form a complete C-type domain.

Domain 2 has two extensions at faces which are equivalent to Q both under $\mathfrak{G}_2$ and under the group of automorphisms of the domain itself. Both extensions of the domain 2 into the domain R_1 and its equivalents are equivalent to the domain $2aaaa00bbb$. Domain 2 and both of its extensions form a complete C-type domain.

In the case of domains 3, 4 and 5 there are two extensions at faces which are equivalent to Q under $\mathfrak{G}_2$. But here the two faces are not mutually equivalent under the group of automorphisms of their respective domains. For this reason each of these two extensions into R_1 or its equivalents are different. The extensions of domain 3 are the domains $3aaaa30bbb$ and $1aaa002bbb$, the extensions of domain 4 are the domains $aaaaa22bbb$ and $2aaa002bbb$, and the extensions of domain 5 are the domains $aaaab22bbb$ and $1aaa000bbb$. This latter was given as the reason for formulating Proposition 8.2.

The remaining domains, 6–21, of R_2 have a single extension at Q. For domains 6–18 these extensions also have only a single extension, and therefore the complete C-type domain corresponding to these consists of only two pieces. As is apparent from the above, domains 21, 20 and 19 correspond to the following C-type domains: 1) the domain $aaaabbbbbb$ with five domains equivalent to 21; 2) the domain $aaaabb0bbb$ with two domains equivalent to domain 20; 3) the domain $aaaaaaabbb$ with two domains equivalent to domain 19.

These investigations show that the complete C-type domains containing each of the domains 1–21 of the perfect domain R_2 are mutually inequivalent. We may now rewrite Propositions 7.4 and 7.5 with more precision:

THEOREM 8.3. *The primitive five-dimensional parallelohedra corresponding to forms from the perfect domain R_2 partition into* 21 *C-types. Each of these C-types corresponds to one of the domains whose diagram is included in Table* II.

On the other hand, of the 24 intersections of C-type domains with R_1 which are mutually inequivalent under $\mathfrak{G}_1$ and which protrude at the face Q (8.8), only 21 are mutually inequivalent under the entire group $\mathfrak{G}$.

Let us turn our attention to the fact that the arguments we use in the proof of Theorem 8.3 may be replaced by more general and shorter lines of reasoning using the material of subsection 1. However, the latter would give less information about the structure of the subdivision of the cone K into C-type domains.

5. We have established that the C-type domains whose intersections with R_1 appear as numbers 1–7 in Table I are all mutually inequivalent, and that the cone

K does not have any other C-type domains inequivalent either to these or to the domain R. This finishes the derivation of the general C-types of the five-dimensional parallelohedra. We have proved

THEOREM A(8.4). *In all there exist exactly* 76 *general geometrical C-type primitive five-dimensional parallelohedra. One of these C-type domains corresponds to the domain R of the first perfect form. The C-type domains corresponding to the remaining* 75 *general C-type parallelohedra intersect the domain* R_1 *of the second perfect form. These intersections are enumerated in diagrams* 1–75 *of Table* I.

CHAPTER II

DERIVATION (AND ENUMERATION) OF ALL GENERAL L-TYPES OF PRIMITIVE FIVE-DIMENSIONAL PARALLELOHEDRA

§9. Decomposition of the C-type domains into L-type domains: General theory; the case $n = 4$.

1. In order to consider the decomposition of a given C-type domain into L-type domains it is necessary to understand the geometry of transformations which carry one primitive L-subdivision into another where the L-type domains corresponding to these two L-subdivisions have in E^N a common $(N-1)$-dimensional face. We will call such transformations, which were first investigated by Voronoĭ, *repartitionings.*

The geometrical interpretation of a theorem of Voronoĭ ([2], §88), lying at the basis of the theory of repartitionings, consists of the following.

The points of the $(N-1)$-dimensional face, at which the repartitioning occurs, correspond to special lattices. The L-subdivisions of these special lattices contain, besides possibly some L-simplices, L-polyhedra (Voronoĭ denoted these bodies by K) having the combinatorial structure of the standard body V^n_{pq} which is described below (subsection 2). On one side of the $(N-1)$-dimensional face where the repartitioning acts, the L-polyhedron has one subdivision into L-simplices; on the other side, it has another. We will call these polyhedra, as well as standard polyhedra combinatorially equivalent to them, *bodies* V^n_{pq}. We will call the two possible subdivisions of V^n_{pq} into L-simplices the *simplicial complexes* of V^n_{pq}. We will call the body V^n_{pq} along with its two simplicial complexes a *repartitioning complex.* We will preserve such names for the standard bodies.

We turn now to a description of the standard body and the simplicial complexes of V^n_{pq}.

2. Let $n = p + q$. In E^n we take two mutually orthogonal flats E^p and E^q intersecting at the origin 0. In E^p we construct a p-dimensional regular simplex S^p, and in E^q we construct a q-dimensional simplex S^q; both have centers at the origin. The standard body V^n_{pq} is defined to be the convex hull of the union of S^p and S^q. We now consider two simplicial subdivisions of V^n_{pq}. The simplices of the first subdivision are the convex hull of S^p and the $(q-1)$-dimensional faces of S^q. In the second subdivision the simplices S^p and S^q change roles.

As in the above we will call these two subdivisions of the standard body V^n_{pq} the (standard) simplicial complexes of V^n_{pq}, specifying, if necessary, which of the two complexes is under consideration. We remark that one of the complexes contains

$1 + p$ simplices of maximal dimensionality; the other, $1 + q$. Let us agree to call S^p and S^q the *basic simplices* of V^n_{pq}.

In the case where $n = p + q + k$ $(k > 0)$ the standard repartitioning complex of V^n_{pq} is constructed in the following manner. In some flat $E^{p+q} \subset E^n$ we take the complex V^{p+q}_{pq} and exterior to this flat we take k points so that the linear span of E^{p+q} and these points coincides with the entire space E^n. Then a simplicial complex of V^n_{pq} is composed of simplices whose vertices form a subset of the chosen points and the vertices of V^{p+q}_{pq}. The repartitioning complex is formed from a system of such simplicial complexes "based" on the common simplicial complex V^{p+q}_{pq}. For example, in the case where $k = 1$, the simplicial complexes are pyramids generated by V^{n-1}_{pq} and have pairwise intersections. We will not need a more detailed description of the complexes V^n_{pq} for the cases $k > 1$.

LEMMA 9.1. *The repartitioning complexes of V^n_{11} are the only ones that occur among the repartitioning complexes of V^n_{1q}. These and only these complexes correspond to transformations from one C-type to another.*

PROOF. If a repartitioning complex of V^{1+q}_{1q}, where $q > 1$, occurred, then one of its simplicial complexes would consist entirely of two L-simplices, and the second of $1 + q$ simplices. Moreover, under a transformation from the second to the first an L-edge would be lost, not being replaced by a new one having the same midpoint.(1) Since such a situation is impossible, the first part of the lemma is proved. The second part follows from the fact that an exchange of one L-edge for another occurs only for the repartitioning complexes of V^n_{11}.

We emphasize that it is the basic sense of Lemma 9.1 that under a repartitioning by means of an arbitrary complex, with the exception of V^n_{11}, the composition of the contiguity vectors of the lattice does not change.

Thus, as we investigate the subdivision of the C-types into L-types, we will not encounter the repartitioning complexes of V^n_{11}; the lattices containing such complexes correspond to faces of C-type domains. It follows that for $n \leqslant 5$ we only need to consider the complexes V^4_{22}, V^5_{22} and V^5_{23}.

3. Recall (§§3 and 7) that if the lattice basis is chosen using transformations (3.3) or (7.2), then the coordinates of lattice vectors must be either whole numbers or irreducible fractions with denominator 2. In the first case the vectors of a lattice, arbitrary segments defined by them, and the L-edges are called respectively integral vectors, integral segments, and integral L-edges. In the second case they are called fractional vectors, segments, and L-edges.

LEMMA 9.2. *Suppose that 6 points of the lattice $\Gamma \subset E^n$ are connected by segments. Then the union of these segments (the 1-dimensional skeleton of the "hexagon") has one of the following four forms*:

1. *All of the segments of the skeleton are integral.*

---

(1) A priori, with the loss of an L-edge one may presuppose the appearance of a new one from another complex V^{1+r}_{1r} $(r > 1)$, but the midpoint of such an L-edge would be different.

2. *Integral segments unite* 5 *vertices*; *the remaining* 5 *segments of the skeleton are fractional.*

3. *Integral segments unite* 4 *vertices and also the two remaining*; *the other* 8 *segments of the skeleton are fractional.*

4. *Integral segments form the sides of two triangles without a common vertex, and the remaining* 9 *segments are fractional* (*see Figure* 10, *in which integral segments correspond to heavy lines and fractional segments to thin ones*).

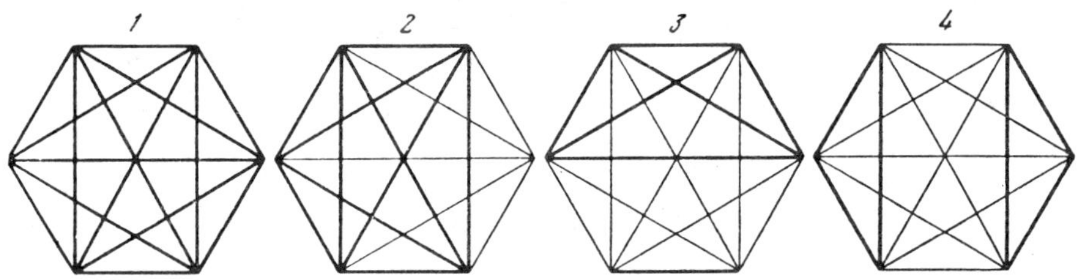

FIGURE 10

PROOF. If the union of the integral segments of the one-dimensional skeleton is disconnected and collapses to several connected components, then there are two and only two components.

In fact, a triangle may appear in the skeleton in one of two ways: either all of its sides are integral, or one is integral and two are fractional. This excludes the existence of one-dimensional skeletons of m vertices with a single fractional segment where $m \geqslant 3$ and (consequently) the existence of three or more components having integral segments.

Let a component having integral segments have $m \geqslant 3$ vertices. Among these, two vertices A_1 and A_2 can be found which are connected by an integral segment. In addition, a vertex A_3 can be found connected to either A_1 or A_2 by an integral segment. But in this case the entire one-dimensional skeleton of the triangle $A_1A_2A_3$ consists of integral segments.

If the component has more than three vertices, then there must be among them a vertex A_4 connected to A_1, A_2 or A_3 by an integral segment. But then the entire one-dimensional skeleton of the figure $A_1A_2A_3A_4$, with four vertices, consists of integral segments.

Repeating these arguments the necessary number of times, we find that fractional segments do not occur and that the vertices of the skeleton are connected by integral segments. It is now obvious that the only possible structures for such a component are those represented in Figure 10. Thus the proof of the lemma is finished.

4. For the continuation of our investigation it is necessary to consider laminar planes of lattices, i.e. $(n-1)$-dimensional planes in E^n which contain faces of polyhedra which appear in the L-subdivision of a lattice $\Gamma \subset E^n$.

LEMMA 9.3. *In order that some* $(n-1)$*-dimensional plane* P *be a laminar plane of a primitive lattice* Γ, *it is necessary and sufficient that the edges of the* L*-subdivision not lying in* P *do not have interior points in* P.

PROOF. Since the lattice Γ is primitive, all of its L-polyhedra are simplices. An L-simplex exists having interior points lying in P if and only if P is not a laminar plane. Therefore interior points of one of the edges of such an L-simplex must lie in the plane. The assertion of the lemma follows from this.

5. We shall consider the body V_{22}^4 and its simplicial complexes in E^4. Let $A_1A_2A_3$ and $B_1B_2B_3$ denote the base triangles of V_{22}^4. From the description of the standard body V_{pq}^n it is easy to deduce in the present case that all 15 segments connecting the vertices $A_1, A_2, A_3, B_1, B_2, B_3$ of V_{22}^4 lie on the surface of this body and consequently are L-edges. Since there is no mutual homology among them (see [10]), they constitute the set of all L-edges for 4-dimensional lattices whose L-subdivisions contain V_{22}^4.

Since the repartitioning complexes of V_{22}^n have as "foundation" the complexes V_{22}^4, by what has been said we have the following lemma:

LEMMA 9.4. *In order that a given C-type be repartitioned by the complex* V_{22}^n *it is necessary that* 15 *distinct vectors be found among its contiguity vectors each of which is parallel to those of a four-dimensional plane and which constitute a one-dimensional skeleton with six vertices. The repartitioning complexes of* V_{22}^4 *on which the complexes of* V_{22}^n *are based are only found among the convex hulls of such families of six vertices.*

We turn to the two inequivalent C-type domains Δ_1 and Δ_2 for the case $n = 4$ (§4) where a repartitioning into L-type domains is a priori possible (we recall that for $n = 4$ there exists only a single C-type domain Δ_0 coinciding with the perfect domain R, and it comprises an L-type domain).

As follows essentially from the results of §4, the list of contiguity vectors for the C-type domains intersecting with the perfect domain R_1 contains 8 fractional vectors and 4 integral vectors which correspond to the pretypes (1, 0, 0, 0), (0, 1, 0, 0), (0, 0, 1, 0) and (0, 0, 0, 1). With the exception of these twelve "constant" vectors there are three additional contiguity vectors which in the case of Δ_1 correspond to the pretypes l^{12}, l^{13} and l^{14} and in the case of Δ_2 correspond to the pretypes $l^{\overline{12}}$, $l^{\overline{13}}$ and $l^{\overline{23}}$.

Below we establish that for $n = 4$ a partitioning of a C-type domain is possible only by means of the complex V_{22}^4. For both inequivalent C-type domains intersecting R_1 we find 8 fractional vectors and 7 integral among the 15 essentially distinct contiguity vectors. According to Lemma 9.2 the one-dimensional skeletons having six vertices constructed from these vectors have form 3 of Figure 10. From this it follows that the existence of such skeletons implies the presence of 6 integral vectors forming the one-dimensional skeleton of a tetrahedron.

In the case of the domain Δ_1 all of the integral contiguity vectors are given by

$$\pm(1,0,0,0),\ldots, \pm(0,0,0,1), \pm(1,1,0,0), \pm(1,0,1,0), \pm(1,0,0,1).$$

As is easy to see, it is impossible to form the one-dimensional skeleton of a tetrahedron from any 6 of these vectors. According to Lemma 9.4 it follows that the C-type domain Δ_1 does not decompose into distinct L-types. Thus Δ_1 corresponds to a single L-type domain (the forms of this domain correspond to those of Voronoĭ's type II domain).

From the contiguity vectors of the domain Δ_2,

$$l^1=(1,\ 0,\ 0,\ 0),\quad l^2=(0,\ 1,\ 0,\ 0),\quad l^3=(0,\ 0,\ 1,\ 0),\quad l^4=(0,\ 0,\ 0,\ 1),$$
$$l^{\overline{12}}=(1,\ -1,\ 0,\ 0),\quad l^{\overline{13}}=(1,\ 0,\ -1,\ 0),\quad l^{\overline{23}}=(0,\ 1,\ -1,\ 0),$$
$$d^0=({}^1/_2,\ {}^1/_2,\ {}^1/_2,\ {}^1/_2),\quad d^1=(-{}^1/_2,\ {}^1/_2,\ {}^1/_2,\ {}^1/_2),\quad d^2=({}^1/_2,\ -{}^1/_2,\ {}^1/_2,\ {}^1/_2),$$
$$d^3=({}^1/_2,\ {}^1/_2,\ -{}^1/_2,\ {}^1/_2),\quad d^4=({}^1/_2,\ {}^1/_2,\ {}^1/_2,\ -{}^1/_2),$$
$$d^{12}=(-{}^1/_2,\ -{}^1/_2,\ {}^1/_2,\ {}^1/_2),\quad d^{13}=(-{}^1/_2,\ {}^1/_2,\ -{}^1/_2,\ {}^1/_2),$$
$$d^{23}=({}^1/_2,\ -{}^1/_2,\ -{}^1/_2,\ {}^1/_2),$$

a one-dimensional skeleton having six vertices may be constructed (Figure 11), which is unique if those homologous to it are not considered. Thus the existence of a simple decomposition is implied and Δ_2 may be partitioned into L-type domains.

We now consider the convex hull of the one-dimensional skeleton of Figure 11. By finding the centers of gravity of the triangles $A_1A_2A_3$ and $B_1B_2B_3$ of the hexagon (Figure 11), we see that they coincide. From this follows the affine and moreover the combinatorial equivalence of the convex hull $V = A_1A_2A_3B_1B_2B_3$ and the body V_{22}^4, where the triangles $A_1A_2A_3$ and $B_1B_2B_3$ play the role of a triangle basis.

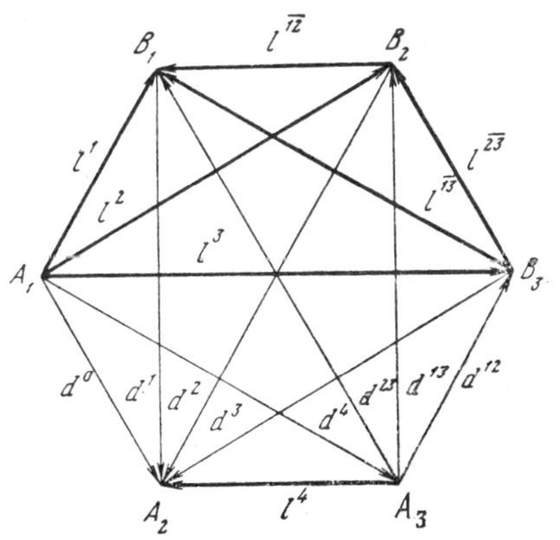

FIGURE 11

We will show that the polytope V is an L-polyhedron, i.e. the union of polytopes of an L-subdivision. This will follow if we can establish that the boundary of V is formed from the three-dimensional faces of L-simplices.

It is easy to see that the boundary of the polytope is formed from 9 tetrahedra $T_{hk} = A_iA_jB_mB_p$, where $h, k = 1, 2, 3$ and the triples h, i, j and k, m, p are permutations of the numbers 1, 2, 3. Since none of the three-dimensional planes defined by these tetrahedra contain interior points of L-edges (of contiguity vectors), by Lemma 9.3 they are all laminar planes. From this it follows that the polytope V is in fact an L-polyhedron.

LEMMA 9.5. *In order that the L-polyhedron V coincide with the body V_{22}^4 (with a basis of triangles $A_1A_2A_3$ and $B_1B_2B_3$) and belong to a special lattice where a repartitioning occurs it is necessary and sufficient that the lengths of its edges satisfy the equality*

$$A_iB_k^2 + A_jB_k^2 + B_mB_p^2 = A_hB_m^2 + A_hB_p^2 + A_iA_j^2, \tag{9.1}$$

where h, $k = 1, 2, 3$ and the triples h, i, j and k, m, p are permutations of the numbers $1, 2, 3$. *When the inequality*

$$A_iB_k^2 + A_jB_k^2 + B_mB_p^2 > A_hB_m^2 + A_hB_p^2 + A_iA_j^2 \tag{9.2}$$

holds the body V_{22}^4 subdivides into three L-simplices of such a form that each has a two-dimensional face coincident with the base triangle $A_1A_2A_3$; when the inequality contrary to (9.2) *holds the triangle $B_1B_2B_3$ is exchanged for the triangle $A_1A_2A_3$, i.e. we have the second simplicial complex.*

PROOF. Consider the simplex $S = A_iA_jB_mB_pC$ whose vertex C is found from the condition $\overline{CA_i} = \overline{A_jB_k}$. We will show that S is an L-simplex of the lattice. First of all we note that the tetrahedron T_{hk} lies on its boundary but does not belong to the polytope V and that all of its edges are L-edges. Since V is an L-polyhedron, there must be an L-polytope contiguous with it at the face T_{hk}. This polytope is an L-simplex, since the lattice cannot be primitive because of the occurence of the repartitioning complex V_{22}^4 and because T_{hk} is not the face of a polytope homologous to V. Making simple choices of L-edges originating from the vertices of the tetrahedron T_{hk}, we conclude that there exists a unique simplex, the edges of which are L-edges, and which is contiguous to the L-polyhedron V at the face T_{hk}. This is our simplex S. Consequently S is an L-simplex of the lattice.

We now choose S as the basic simplex for a barycentric system of coordinates $x_0, x_1, \dots, x_4$. We assume that $C = (1, 0, 0, 0, 0)$, $A_i = (0, 1, 0, 0, 0)$, $A_j = (0, 0, 1, 0, 0)$, $B_m = (0, 0, 0, 1, 0)$ and $B_p = (0, 0, 0, 0, 1)$. In this case the barycentric coordinates x_1, x_2, x_3, x_4 coincide with our usual coordinates. But the point B_k will have the barycentric coordinates $(-1, 1, 1, 0, 0)$, and the point A_h the coordinates $(-1, 0, 0, 1, 1)$.

According to [34], Theorem 2, the basic L-simplex S' of the lattice contiguous to the L-simplex S at the face T_{hk} has the property that five of its vertices correspond to points of the lattice of the form $(-1, x_1, x_2, x_3, x_4)$, where the coordinates x_1, x_2, x_3, x_4 are either permutations of the numbers $(1, 1, 0, 0)$ or permutations of the numbers $(1, 1, 1, -1)$. Thus these vertices correspond to either A_h or B_k.

We denote by h_{rs} the square of the length of an edge of the basic simplex S, the ends of which correspond to vectors having r and s coordinates equal to one, and by $\Phi(X)$ the following function of the coordinates $x_0, x_1, \dots, x_4$:

$$\Phi(X) = \Phi(x_0, x_1, \dots, x_4) = -\sum_{\substack{r,s=0\\(r<s)}}^{4} h_{rs}x_rx_s. \tag{9.3}$$

Then, on the strength of a lemma in [34], the point A_h will be a vertex of S' if the inequality

$$\Phi(A_h) < \Phi(B_k) \tag{9.4}$$

holds, and B_k willl be one of its vertices if the opposite inequality holds. By the same lemma, if

$$\Phi(A_h) = \Phi(B_k), \tag{9.5}$$

then we have a special lattice, the L-subdivision of which contains V.

Substituting the coordinates of A_h and B_k into (9.4) yields

$$h_{03} + h_{04} - h_{34} < h_{01} + h_{02} - h_{12},$$

or

$$A_hB_m^2 + A_hB_p^2 - B_mB_p^2 < A_iB_k^2 + A_jB_k^2 - A_iA_j^2,$$

which is equivalent to (9.2). Also (9.5) is equivalent to (9.1).

By the same token we have shown that (9.1) ($h, k = 1, 2, 3; h \neq k$) holds if and only if V is an L-polyhedron, and that (9.2) or its opposite holds if and only if V is one of the two simplicial complexes of V_{22}^4. This finishes the proof of Lemma 9.5.

We investigate repartitionings realized in the domain Δ_2 by means of the complexes V_{22}^4 (the polyhedron V) which we have analyzed. Substituting into (3.4) the coordinates of the contiguity vectors corresponding to L-edges the squares of the lengths of which appear in (9.1) (and (9.2)), we find the squares of the lengths of these edges. Then we substitute these lengths into (9.1) and (9.2). Independently of the value of the indices, we obtain that each of the equalities (9.1) is equivalent to

$$\rho_{\overline{12}} + \rho_{\overline{13}} + \rho_{\overline{23}} = 0, \tag{9.6}$$

and each of the inequalities (9.2) is equivalent to

$$\rho_{\overline{12}} + \rho_{\overline{13}} + \rho_{\overline{23}} > 0. \tag{9.7}$$

Obviously, the inequalities which are opposite to (9.2) are equivalent to the inequality which is opposite to (9.7). By the same token we learn that (9.6) defines the set of points of Δ_2 where this C-type domain subdivides into L-type domains.

As was established above (§§3 and 4), equation (9.6) defines in E^N ($N = 10$) a nine-dimensional plane which contains one of the faces of R_1 separating it from a domain R_1' which is equivalent to it. The C-type domain Δ_2 is subdivided into two domains by this face. Since there are no other repartitioning complexes for lattices of Δ_2, these two domains are L-type domains. The inequality which is opposite to (9.7) shows that the corresponding simplicial subdivision of V_{22}^4 corresponds to lattices which are exterior to R_1.

The transformation

$$x_1 \to x_1 + \frac{x_2 + x_3}{2}, \qquad x_2 \to -x_4 - \frac{x_2 + x_3}{2},$$

$$x_3 \to x_4 - \frac{x_2 + x_3}{2}, \qquad x_4 \to \frac{x_3 - x_2}{2}$$

maps the perfect domain R_1' into the perfect domain R_1 and the L-type domain $\Delta_2 \cap R_1'$ into the L-type domain $\Delta_2 \cap R_1$. This establishes the equivalence of the two

L-type domains into which Δ_2 subdivides. Consequently Δ_2 defines a unique four-dimensional L-type parallelohedron; this is Voronoĭ's type III.

Thus we have proved the following theorem.

THEOREM 9.6. *For $n = 4$, of the three inequivalent C-type domains Δ_0, Δ_1 and Δ_2, the two domains Δ_0 and Δ_1 each coincide with a single L-type domain and Δ_2 subdivides into two L-type domains, the boundary of this subdivision coinciding with a face of the perfect domain R_1. The two L-type domains into which Δ_2 subdivides are equivalent, and by the same token they determine one and only one L-type four-dimensional parallelohedron* (*type* III *of Voronoĭ*).

The well-known fact that for $n = 4$ the subdivision of the cone K into L-types is a subdecomposition of the perfect subdivision of this cone follows directly from Theorem 9.6.

We remark that a unique special type lies on the face $\rho_{\overline{12}} + \rho_{\overline{13}} + \rho_{\overline{23}} = 0$ which subdivides Δ_2 into L-type domains. Astonishingly enough, the four-dimensional parallelohedra (see [10], Figure 45, No. 19) corresponding to these have exactly 30 faces, as do the primitive parallelohedrons. The existence and uniqueness of such degenerate types of Voronoĭ-Dirichlet domains for $n = 4$ is now easily explained by the existence and uniqueness of the partitioning plane for the C-type domain Δ_2.

§10. L-subdivisions containing L-simplices of double volume and the L-type primitive parallelohedra that correspond to them (the L-types of the domain R_2)

1. In the work of Voronoĭ ([2], §67) we have mention of the fact that primitive parallelohedra which correspond to L-subdivisions containing nonbasic simplices are found in five-dimensional space, i.e., simplices whose volume is some integral multiple of that of a basic simplex of the lattice. These and only these n-dimensional parallelohedra can have fewer than $(n + 1)!$ vertices. There is no additional explicit data bearing on these parallelohedra appearing in any of the published work of Voronoĭ.

The first example of a primitive n-dimensional lattice ($n \geqslant 5$) whose L-subdivision contains a nonbasic simplex appeared in [35] and [36]. In particular, in these works it was shown that for E^5 there were primitive lattices which had L-simplices of double volume, i.e. of volume $2v$, where v is the volume of a basic complex of the lattice. Further, it was established in [37] that the L-subdivisions of five-dimensional lattices do not contain nonbasic simplices other than those of volume $2v$.

In this section we will find all L-type primitive five-dimensional parallelohedra which correspond to L-subdivisions containing simplices of double volume. At the same time, as will become clear, we find all of the L-types of the domain R_2.

We denote a simplex of the lattice $\Gamma \subset E^5$ by S, i.e. a simplex with vertices at the lattice points $A_0, A_1, \ldots, A_5$. Let $h_{pq} = h_{qp} = A_pA_q^2$ ($p, q = 0, 1, \ldots, 5$; $p \neq q$). We will utilize the following propositions which in fact were given in the note [37].

PROPOSITION 10.1. *In order that the simplex S of double volume be an L-simplex of the lattice, it is necessary and sufficient that* (a) *its four-dimensional faces be basic simplices of their lattices, and* (b) *the square of the lengths of its edges satisfy the inequality*

$$\sum_{\substack{r=0\\(r\neq p\neq q)}}^{5}(h_{pr}+h_{qr})>h_{pq}+\sum_{\substack{r,\,s=0\\(r\neq s\neq p\neq q,\ r<s)}}^{5}h_{rs}\qquad(p,\ q=0,\ 1,\dots,5;\ \ p<q). \tag{10.1}$$

PROPOSITION 10.2. *Let the edges of some simplex satisfy condition* (10.1). *Then there exists a unique lattice for which this simplex is an L-simplex of double volume.*

Let the simplex S be a simplex of double volume of the lattice satisfying condition (a) of Proposition 10.1. Then the quantities $h_{01}, h_{02}, \dots, h_{45}$ may be considered as parameters of the lattice. In the space of these parameters the inequalities (10.1) define an infinite simplicial pyramid Σ. From Proposition 10.2 it follows that interior points of Σ, and only these points, correspond to lattices constructed from L-simplices of double volume.

THEOREM 10.3. *All interior points of the domain R_2 correspond to coordinate frames of lattices with L-simplices of double volume. The remaining part belongs to the domain of reduction $\mathfrak{R}$, which is free of such points.*

PROOF. Simplices constructed with coordinate frames of lattices corresponding to the forms (7.3) have double volume, since the determinant of the transformation (7.2) is equal to 2. Calculating the squares of the lengths of the edges of such simplices, assuming $A_0 = (0, 0, \dots, 0)$, $A_1 = (1, 0, \dots, 0), \dots, A_5 = (0, 0, \dots, 1)$, we find the values h_{pq} $(p, q = 0, 1, \dots, 5; p < q)$. By direct calculation or by taking advantage of the proof of Theorem 7.1 we get

$$h_{pq}=f(\overline{l^{pq}})=\sum_{r=0}^{5}(\sigma_{pr}+\sigma_{qr})\qquad(p,\ q=0,\ 1,\dots,5;\ \ p<q), \tag{10.2}$$

where $\overline{l^{0i}} = l^i$ $(i = 1, \dots, 5)$ and the coordinates of the vectors l^i and $\overline{l^{ij}}$ $(i, j = 1, \dots, 5; i < j)$ are defined in §7. Substitution of the values (10.2) into the left- and right-hand sides of (10.1) shows that these inequalities are equivalent to the following:

$$8\sigma_{pq}>0\qquad(p,\ q=0,\ 1,\dots,5;\ \ p<q). \tag{10.3}$$

According to (7.3), inequalities (10.3) define the set of all interior points of R_2. Condition (10.1), in the above form, is satisfied for all simplices of double volume constructed using coordinate frames which correspond to points of R_2. Since condition (a) of Proposition 10.1 is fulfilled for these, the first part of the theorem follows from this proposition.

We will show that the mapping of R_2 into the domain Σ given by the correspondence (10.2) is one-to-one. In fact, lattices whose parameters h_{pq} satisfy (10.1) have frames $(\overline{A_0A_1}, \overline{A_0A_2}, \ldots, \overline{A_0A_5})$. The quadratic forms corresponding to these frames are contained in the interior of R_2, i.e. each point $(h_{01}, \ldots, h_{45}) \in \Sigma$ corresponds to a point $f \in R_2$, and all of the simplices of Σ map into R_2 (if the mapping is chosen in a different way, Σ will go over into a domain which is equivalent to R_2).

From the one-to-oneness of the transformation based on the correspondence (10.2) between the domains R_2 and Σ, and from the fact that $\mathfrak{R}\backslash R_2$ does not contain forms equivalent to forms of R_2, the second part of the theorem follows.

Since some of the inequalities (10.3) become equalities on the boundary of R_2, it follows that lattices corresponding to these points become special lattices, and the simplex $S = A_0A_1 \cdots A_5$ is no longer an L-simplex. From this fact and from Theorem 10.3 we get

THEOREM 10.4. *The perfect domain R_2 is subdivided into L-type domains.*

We will establish one more fact regarding the L-subdivision of lattices containing simplices of double volume.

PROPOSITION 10.5. *Among the mutually nonhomologous simplices in the L-subdivision of a five-dimensional lattice there is at most one simplex of double volume.*

PROOF. Let S be a nonsymmetrical L-simplex of double volume of a primitive five-dimensional lattice. Corresponding to the possible orderings of the vertices A_0, $A_1, \ldots, A_5$, of the simplex we obtain $h^{(1)}, h^{(2)}, \ldots, h^{(6!)}$ distinct points in the space of parameters h_{pq} lying in the interior of Σ. On the basis of the mapping of Σ into R_2 described in the proof of Theorem 10.3, the points $h^{(1)}, h^{(2)}, \ldots, h^{(6!)}$ correspond to forms $f^{(1)}, f^{(2)}, \ldots, f^{(6!)}$ which are points interior to R_2. Obviously these forms are mutually equivalent under the group $\mathfrak{G}_2$ of automorphisms of R_2.

Since the number of forms $f^{(1)}, f^{(2)}, \ldots, f^{(6!)}$ coincides with the number of elements of $\mathfrak{G}_2$, forms $f', f'' \in R_2$ corresponding to nonsymmetrical, noncongruent L-simplices S' and S'' of double volume cannot be equivalent under $\mathfrak{G}_2$. They will also not be equivalent under the group $\mathfrak{G}$ of automorphisms of the cone K. From this it follows that S' and S'' cannot belong to the same lattice. Thus the theorem is proved in the case of lattices with nonsymmetrical simplices S.

The above argument extends trivially to the case of symmetrical simplices S.

2. We will consider repartitioning complexes for the case $n = 5$. It was established in §9 that in addition to V_{11}^5 we will only need to consider the complexes V_{22}^5 and V_{23}^5.

LEMMA 10.6. *In the case of a repartitioning by the complex V_{22}^5 the body V_{22}^5 under any of the simplicial L-subdivisions subdivides into six basic simplices of the lattice. Under a repartitioning by the complex V_{23}^5 the body V_{23}^5 subdivides into four basic simplices under one L-subdivision and subdivides into a single simplex of double volume and two basic simplices under the other.*

PROOF. The body V_{22}^5 is composed of two pyramids with bases consisting of the body V_{22}^4. Let $A_1 A_2 A_3$ and $B_1 B_2 B_3$ be the basic triangles of V_{22}^4. Then one of its simplicial subdivisions consists of the simplices $A_1 A_2 A_3 B_i B_j$ ($i, j = 1, 2, 3$; $i < j$); the other, of the simplices $B_1 B_2 B_3 A_i A_j$ ($i, j = 1, 2, 3; i < j$). In the first and second cases V_{22}^4 subdivides into three L-simplices of the four-dimensional lattice containing V_{22}^4. These simplices, as with all L-simplices of four-dimensional lattices, are basic.

The pyramids of V_{22}^5 which are generated by the simplicial complex V_{22}^4 each subdivide into three five-dimensional L-simplices having the same volume but which are not homologous. According to Proposition 10.5, these simplices cannot have double volume. Consequently they are basic, and the first part of the lemma is proved.

The body V_{23}^5 (see §9.2) is the convex hull of the union of the basic triangle $A_1 A_2 A_3 \subset E_1^2$ and the basic tetrahedron $B_1 B_2 B_3 B_4 \subset E_2^3$, $E_1^2 \cap E_2^3 = \{0\}$. One of the simplicial complexes V_{23}^5 consists of the four L-simplices $A_1 A_2 A_3 B_i B_j B_k$ ($i, j, k = 1, 2, 3, 4; i < j < k$); the other consists of the three L-simplices $B_1 B_2 B_3 B_4 A_i A_j$ ($i, j = 1, 2, 3; i < j$). Since the number of simplices of the second complex is one greater than that of the first, the first must contain at least one nonbasic simplex. In both complexes the simplices are mutually nonhomologous. By Proposition 10.5 and the fact that no L-simplices have greater than double volume in E^5 we deduce that the sum of volumes of the simplices generating the complex V_{23}^5 is exactly $4V$. From this we obtain the second assertion of the lemma.

LEMMA 10.7. *The boundary of the domain R_2 consists of the set of forms of $\mathfrak{R}$ where a repartitioning by the complex V_{23}^5 occurs. All other repartitionings of C-type domains into L-type domains (in $\mathfrak{R}$) proceed by means of the complex V_{22}^5.*

PROOF. Earlier in this section we established that the boundary of R_2 in the perfect domain of reduction $\mathfrak{R}$ consists of the set of forms where a transition occurs from an L-subdivision containing simplices of double volume to an L-subdivision not containing these simplices. Consequently, the boundary of R_2 consists of those forms where a repartitioning by the complex V_{23}^5 occurs.

Assume that the subdivision based on the fixed complex V_{23}^5 occurs at a set ω of interior points of R_2. We can establish a correspondence between the points of ω and neighborhoods of values of the parameters h_{pq} of these points, using a coordinate frame whose vectors are parallel to the edges of the L-simplex of double volume from our complex V_{23}^5. This correspondence degenerates at the points of ω. By means of this correspondence ω is represented by the set σ of boundary points of Σ. Then, with the help of the one-to-one map (10.2), the set σ is mapped into some set ω' of boundary points of R_2. In this way we have obtained a transformation of the group $\mathfrak{G}$ mapping a set ω of interior points of R_2 into a set ω' of boundary points, which is impossible.

At interior points of R_1, the appearance (or disappearance) of simplices of double volume is impossible, and therefore a repartitioning by means of the complex V_{23}^5 cannot occur. Lemma 10.7 follows from this and from the fact that repartitioning complexes other than V_{23}^5 and V_{23}^5 cannot be interior to C-type domains for $n = 5$.

3. We will consider the forms of the domain R_2. In the case of five-dimensional lattices corresponding to these forms the L-subdivision can contain the complex V_{22}^5 only if it contains the four-dimensional body V_{22}^4. Thus for the investigation of V_{22}^5 it is necessary, according to Proposition 9.4, to find four-dimensional planes containing 15 essentially distinct contiguity vectors forming one of the one-dimensional skeletons enumerated in Lemma 9.2.

LEMMA 10.8. *If the body V_{22}^4 is contained in the L-subdivision of a form* (7.3) *of the domain R_2, then the one-dimensional skeleton of V_{22}^4 is contained in the plane*

$$x_p + x_q = 0 \quad \left(p,\ q = 0,\ 1, \ldots, 5;\ \ p \neq q;\ \ x_0 = -\sum_{i=1}^{5} x_i\right) \tag{10.4}$$

and has the form 3 *of Figure* 10.

PROOF. From Theorem 7.1 it follows that the 15 essentially distinct integer contiguity vectors are identically for all forms of R_2. On the basis of the results of subsection 1 and the present subsection we find that these vectors form the edges of an L-simplex of double volume. The one-dimensional skeleton of this simplex is represented in Figure 12. Since this skeleton, the unique skeleton of form 1, is

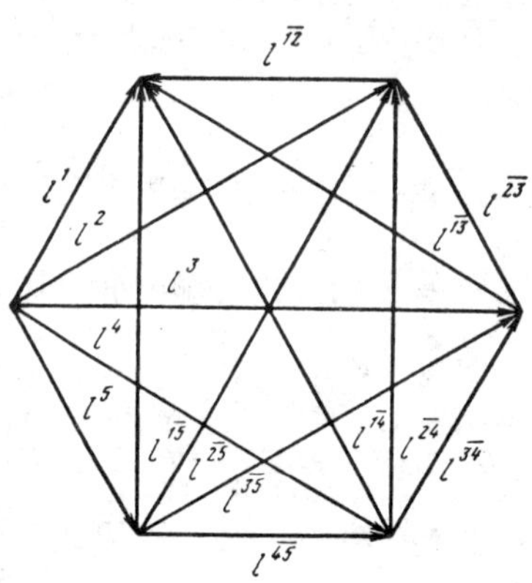

FIGURE 12

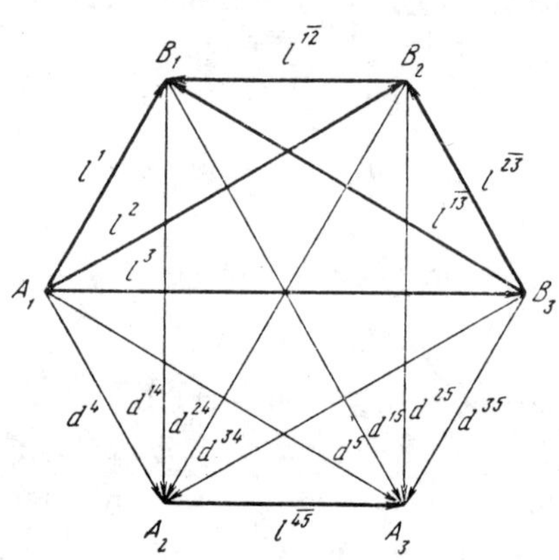

FIGURE 13

formed from the essentially distinct contiguity vectors of lattices of R_2, it follows that the one-dimensional skeleton of V_{22}^4 cannot have form 1.

If we assume that the one-dimensional skeleton of V_{22}^4 has form 2 or form 4, an inspection of the possible integer contiguity vectors of such a skeleton shows that they lie in one of the planes

$$x_p = 0, \quad x_p + x_q + x_r = 0 \quad (p,\ q,\ r = 0,\ 1, \ldots, 5;\ \ p \neq q \neq r). \tag{10.5}$$

Since one of the fractional contiguity vectors of lattices of R_2 does not belong to any of the planes (10.5), the one-dimensional skeleton of V_{22}^4 cannot have either form 2 or form 4.

We consider now the one-dimensional skeletons of form 3 with edges composed of essentially distinct contiguity vectors. It is easy to see that the integer vectors of such a skeleton belong to one of the planes (10.4) and not to any others. The lemma is proved.

THEOREM 10.9. *The intersection of a C-type domain with the domain R_2 is an L-type domain. The domain R_2 contains* 21 *mutually inequivalent L-types* (*Table* II) *and at the same time it defines* 21 *types of five-dimensional primitive parallelohedra.*

PROOF. From Lemmas 10.7 and 10.8 it follows that the intersection of a C-type domain with R_2 will decompose into L-type domains only if the one-dimensional skeleton of V_{22}^4 belongs to the plane (10.4).

All of the planes (10.4) are equivalent under the group $\mathfrak{G}_2$, and therefore, without loss of generality, it is sufficient to consider the plane

$$x_4 + x_5 = 0. \tag{10.6}$$

Besides the integral vectors l^1, l^2, l^3, $l^{\overline{12}}$, $l^{\overline{13}}$, $l^{\overline{23}}$ and $l^{\overline{45}}$ the plane (10.6) contains the constant fractional vectors d^{14}, d^{15}, d^{24}, d^{25}, d^{34} and d^{35}, and the variable fractional vectors d^4 and d^5.

By considering the coordinates of the 15 contiguity vectors which we have enumerated it is easy to deduce that they form a hexagonal one-dimensional skeleton in a unique way (Figure 13).

Suppose that the convex hull V of the one-dimensional skeleton $A_1A_2A_3B_1B_2B_3$ of Figure 13 is the body V_{22}^4. Then the role of triangle basis is played by the triangles $A_1A_2A_3$ and $B_1B_2B_3$.

Calculating the square of the length of the edges of the polyhedron V using (7.3), we see that (9.1) is equivalent to the equality

$$\sigma_{04} + \sigma_{05} = 0, \tag{10.7}$$

and (9.2) is equivalent to the inequality $\sigma_{04} + \sigma_{05} > 0$. Since the parameters σ_{pq} ($p, q = 0, 1, \ldots, 5; p < q$) of forms (7.3) of the domain R_2 are nonnegative, (10.7) implies that $\sigma_{04} = \sigma_{05} = 0$. Since such parameters are equal to zero only on faces of R_2, it follows that forms which satisfy (10.7) lie on a 13-dimensional face of R_2.

Thus we have established that repartitionings by the complex V_{22}^4 at a four-dimensional L-plane do not occur at interior points of R_2. From this it follows that interior to R_2 there are no repartitionings by the complex V_{22}^5, and that the intersection of a C-type domain with R_2 does not decompose into L-type domains. On the strength of Lemma 10.7 we have completed the proof of the theorem.

§11. The complexes of V_{22}^5 in the L-subdivision of lattices corresponding to forms of the domain R_1

1. The first step in the consideration of the repartitioning complexes V_{22}^5 for the L-subdivisions of lattices of R_1 is the determination, using Lemmas 9.4 and 9.2, of the one-dimensional skeleton of the body V_{22}^4.

We recall (§3) that all C-type domains intersecting R_1 belong to 21 fixed pretypes. These pretypes correspond to the following distinct contiguity vectors, which we have agreed above to call constant vectors. The first 16 are fractional vectors: $d^0 = (½, ½, ½, ½, ½)$ and the vectors d^i and d^{ij} $(i, j = 1, \ldots, 5; i < j)$ where $x_i = x_j = -½$, the remaining coordinates being ½. The other 5 are integer vectors l^i $(i = 1, \ldots, 5)$ where $x_i = 1$, the remaining coordinates being 0.

The remaining contiguity vectors are variable, the collection varying for the different C-type domains. For each C-type we choose the 10 essentially distinct variable vectors in such a way that their coordinates coincide with those of the corresponding pretypes. Thus we can make use of the vectors which we have denoted by l^{ij}, $\overline{l^{ij}}$ and $l^{p,ij}$ $(i, j = 1, \ldots, 5; i < j; p = 0, 1, 2, 3)$ for the pretypes (5.1) of §5.

LEMMA 11.1. *All four-dimensional planes containing* 15 *essentially distinct contiguity vectors of an arbitrary C-type of* R_1 *are parallel to one of the planes defined by the equations*

$$x_i \pm x_j = 0 \qquad (i, j = 1, \ldots, 5; \quad i \neq j). \tag{11.1}$$

Among the 15 *essentially distinct contiguity vectors in such a four-dimensional plane there are* 8 (*and only* 8) *fractional vectors.*

PROOF. Let the plane containing the 15 contiguity vectors required by the lemma have the form

$$\sum_{i=1}^{5} a_i x_i = 0. \tag{11.2}$$

Assume that none of the coefficients of (11.2) are equal to 0, i.e. none of the vectors l^i $(i = 1, \ldots, 5)$ belong to this plane. Consequently there are at least 5 essentially distinct fractional vectors $d_1, \ldots, d_q$ $(q \geqslant 5)$ lying in it.

Assume that the vector d_1 has the coordinates $(\delta_1, \ldots, \delta_5)$, where $\delta_i = \pm ½$ $(i = 1, \ldots, 5)$. Then d_2 has the form $(\delta_1, \delta_2, \delta_3, -\delta_4, -\delta_5)$. In fact, if two fractional vectors differ in one (the ith) or in four (jth, kth, mth, pth) coordinates, then one of the coefficients (a_i) of (11.2) must be zero.

The coordinates x_4 and x_5 of the vector d_3 may not coincide with those of d_1 or d_2; otherwise one of the pairs (d_1, d_3) or (d_2, d_3) would force one of the coefficients a_1, a_2 or a_3 to be zero. In such a way, to within an indexing of the coordinates of the triple x_1, x_2, x_3 and the pair x_4, x_5 the vector d_3 has the coordinates $(\delta_1, \delta_2, -\delta_3, \delta_4, -\delta_5)$.

The vector d_4, because of the premise regarding the coefficients $a_1, \ldots, a_5$, must have (to within an indexing of the pair x_1, x_2) either the form $(\delta_1, -\delta_2, \delta_3, \delta_4, -\delta_5)$ or the form $(\delta_1, \delta_2, -\delta_3, -\delta_4, \delta_5)$. It is easy to verify that the second variant of d_4 implies the equality $\delta_3 a_3 - \delta_5 a_5 = 0$, which is incompatible when $a_3 \neq 0$ and $a_5 \neq 0$ with the equality $\delta_3 a_3 + \delta_5 a_5 = 0$ which follows from the pair (d_1, d_3). Thus we have that $d_4 = (\delta_1, -\delta_2, \delta_3, \delta_4, -\delta_5)$.

Analogously we find that $d_5 = (-\delta_1, \delta_2, \delta_3, \delta_4, -\delta_5)$. Substitution of the coordinates of the vectors $d_1, \ldots, d_5$ into (11.2) yields the system of equalities

$$\delta_1 a_1 = \delta_2 a_2 = \delta_3 a_3 = \delta_4 a_4 = -\delta_5 a_5, \qquad \sum_{i=1}^{5} \delta_i a_i = 0,$$

from which it follows that $a_1 = a_2 = \cdots = a_5 = 0$. But this contradicts the hypothesis that none of the coefficients of (11.2) are equal to zero.

Assume that one of the coefficients, for example a_5, is zero. Using the same arguments as above, we can show that in this case there are no more than 6 fractional vectors belonging to the plane (11.2). These vectors, up to a permutation of the coordinates, are the following: $(\delta_1, \delta_2, \delta_3, \delta_4, \pm\delta_5)$, $(\delta_1, \delta_2, -\delta_3, -\delta_4, \pm\delta_5)$, $(\delta_1, -\delta_2, -\delta_3, \delta_4, \pm\delta_5)$, where $\delta_i = \pm\frac{1}{2}$ $(i = 1, \ldots, 5)$. Since of the 5 constant integral vectors only one lies in the plane, no fewer than 8 variable integral vectors must lie in it. But the latter is impossible.

Consider the contiguity vectors of index $J_{15}, J_{25}, J_{35}, J_{45}$. If the pretypes of these four indices or any three of them do not have simple weights, then correspondingly the 4 or 3 contiguity vectors do not belong in the plane (11.2), and the hypothesis is rejected.

Let the pretype of some index, for example J_{15}, be composite. Then the pretypes of indices J_{23}, J_{24}, J_{34} are simple, as is easy to see. In this case there are no contiguity vectors of index J_{15} and at most a single contiguity vector of index J_{23}, J_{24} or J_{34} lying in the plane (11.2). In fact, two contiguity vectors having indices J_{23}, J_{24} or J_{34} lying in the plane (11.2) would imply the relation $|a_2| = |a_3| = |a_4|$. But in the case of a composite pretype this implies that no contiguity vector of index J_{15} belongs to the plane (11.2).

Suppose that two of the four pretypes of index $J_{15}, J_{25}, J_{35}, J_{45}$ have simple weights. Then for the proof it is sufficient to adapt the previous argument for a single pretype of composite weight. In the case of only a single simple weight the argument for two pretypes of composite weight can be adapted, and in the case of no simple weights the argument for three pretypes of composite weight can be adapted.

Let us assume that only two of the coefficients of the equation (11.2) are equal to zero, for example $a_4 = a_5 = 0$. In this case the plane (11.2) contains the two constant integral vectors l^4 and l^5. If it contains fractional vectors, then it contains exactly four of them: $(\delta_1, \delta_2, \delta_3, \pm\frac{1}{2}, \pm\frac{1}{2})$, where the δ_t $(t = 1, 2, 3)$ are fixed. Thus there must be no fewer than 9 variable vectors lying in this plane. That this later situation is impossible follows from the consideration of the contiguity vectors of index $J_{14}, J_{15}, J_{24}, J_{25}, J_{34}, J_{35}$.

If we assume that exactly four of the coefficients in (11.2) are equal to zero, then the plane under consideration will not contain a single fractional contiguity vector. Consequently exactly three coefficients are equal to zero in (11.2), and it has the form

$$a_i x_i + a_j x_j = 0. \tag{11.3}$$

In order that a single fractional vector belong to the plane (11.3) it is necessary and sufficient that the condition $a_i = \pm a_j$ be fulfilled; and if this condition holds then the plane (11.3) will be parallel to 8 (and only 8) fractional contiguity vectors. This proves Lemma 11.1.

COROLLARY. *All four-dimensional planes containing L-edges which are elements of a one-dimensional skeleton with six vertices are parallel to one of the planes* (11.1). *Among the L-edges of this skeleton there are* 8 (*and only* 8) *fractional vectors, i.e. the skeleton has form* 3.

Since a one-dimensional skeleton having four dimensions and six vertices is formed from 15 segments which correspond to essentially distinct contiguity vectors, the corollary is obvious.

This corollary reduces our search for one-dimensional skeletons of the body V^4_{22} to the investigation of collections of L-edges which are parallel to any of the planes (11.1). Since all of the planes (11.1) are equivalent with respect to the group G_1, it is sufficient to consider the case of the plane

$$x_4 + x_5 = 0. \tag{11.4}$$

The plane (11.4) is parallel to 11 constant contiguity vectors, i.e. the eight fractional vectors d^4, d^5, d^{14}, d^{15}, d^{24}, d^{25}, d^{34}, d^{35} and the three integral vectors l^1, l^2, l^3. The edges defined by these 11 vectors necessarily occur in the unknown one-dimensional skeleton of V^4_{22}. It is easy to establish that the plane (11.4) is parallel to only those variable vectors which correspond to pretypes of index (1, 1, 0, 0, 0), (1, 0, 1, 0, 0), (0, 1, 1, 0, 0) and (0, 0, 0, 1, 1).

LEMMA 11.2. *There are two* (*and only two*) *one-dimensional skeletons having six vertices which are mutually inequivalent with respect to the group* G_1, *which are parallel to the plane* (11.4) *and which are formed from L-edges. Besides the* 11 *L-edges defined by the constant vectors, the first of these contains the L-edges which correspond to the vectors* $l^{\overline{45}} = (0, 0, 0, 1, -1)$, $l^{\overline{12}} = (1, -1, 0, 0, 0)$, $l^{\overline{13}} = (1, 0, -1, 0, 0)$ *and* $l^{\overline{23}} = (0, 1, -1, 0, 0)$ (*Figure* 13). *The second differs from the first only in the last two vectors: they are* $l^{3,13} = (0, 1, 0, 1, -1)$ *and* $l^{3,23} = (1, 0, 0, 1, -1)$ (*Figure* 14).

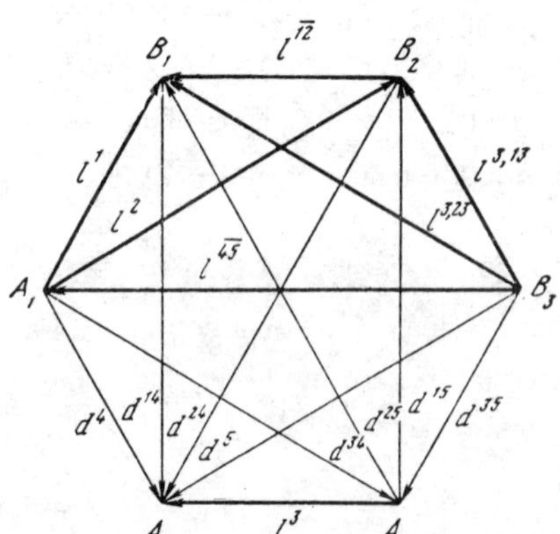

FIGURE 14

PROOF. The desired one-dimensional skeleton has the form 3. For a skeleton of this form there are two integer components. The "large" one represents the one-dimensional skeleton of a tetrahedron and the "small" one consists of a single edge. The 8 fractional edges of the skeleton form four pairs, each of which together with the "small" component with an integer edge yields a triangle. By considering such a partitioning of the 8 vectors d^4, d^5, d^{14}, d^{15}, d^{24}, d^{25}, d^{34}, d^{35} defining fractional edges of the skeleton into pairs we conclude that only the vectors l^1, l^2, l^3 and $l^{\overline{45}}$ can occur in the small component.

Thus there are two inequivalent variants possible: 1) the edges defined by the vectors l^1, l^2 and l^3 all appear in the large component and the small component is defined by the vector $l^{\overline{45}}$; 2) one of the three vectors l^1, l^2 or l^3 (say l^3) defines the small component. Both variants consisting of fractional edges and the small component are uniquely augmented by the integer vectors forming the entire skeleton. The first variant gives the skeleton of Figure 13, and the second variant gives the skeleton of Figure 14. The lemma is proved.

The one-dimensional skeletons equivalent to the one-dimensional skeleton of Figure 13 with respect to the group G_1 we will call *skeletons of form* $\mathfrak{A}$. Those equivalent to the skeleton of Figure 14 we will call *skeletons of form* $\mathfrak{B}$.

2. For our development we will need the information given below regarding the laminar planes of five-dimensional lattices.

LEMMA 11.3. *Let π be one of the planes from the set* (11.1). *In order that π be a laminar plane of a lattice of a given C-type it is necessary and sufficient that of the 7 vectors which are images of the vectors*

$$l^{45},\quad l^{0,12},\quad l^{0,13},\quad l^{0,23},\quad l^{1,12},\quad l^{1,13},\quad l^{1,23} \tag{11.5}$$

under the integral unimodular transformation of the variables $x_1, \ldots, x_5$ which transforms the plane (11.4) *into π, none are included among the variable contiguity vectors of this C-type.*

PROOF. Consider the plane (11.4). Suppose that the origin for the constant vectors and the vectors l^{ij}, $l^{\overline{ij}}$ and $l^{p,ij}$ ($i, j = 1, \ldots, 5; i<j; p = 0, 1, 2, 3$), among which are found the variable contiguity vectors of a lattice of a given C-type, is placed at a point in the plane (11.4). Then the ends of these vectors will lie either in the same plane (11.4) or in one of the planes

$$x_4+x_5=\pm 1, \tag{11.6}$$

and the seven vectors (11.5) will belong to the plane

$$x_4+x_5=2. \tag{11.7}$$

Since there are no lattice points between the plane (11.4) and each of the planes (11.6) which are parallel to it, and since there are lattice points lying between the planes (11.4) and (11.7), it follows from Lemma 9.3 that the plane (11.4) is laminar if and only if none of the vectors (11.5) occur among the variable contiguity vectors of the given C-type.

Since any of the planes (11.1) will transform into the plane (11.4) by some integral unimodular transformation of the variables $x_1, \ldots, x_5$, the lemma is proved.

3. We will call the union of the red edges 12, 13, 23 and 45 of a C-type diagram of R_1 the *α-component,* and we will call the union of the red edge 12, the black edge 45 and the two b-triangles 145 and 245 with black sides the *β-component.* All collections of elements of a diagram which are equivalent under G_1 to either the α- or

the β-component are called the *components* of the diagram.

LEMMA 11.4. *All C-type diagrams from Table* I *which contain a component* κ *equivalent to the* α*-component, and only these diagrams, correspond to C-type domains whose lattices have L-subdivisions which contain the four-dimensional body* V_{22}^4 *with a one-dimensional skeleton of the form* $\mathfrak{A}$.

Let a C-type diagram be chosen such that $\kappa \equiv \alpha$. *Then the four-dimensional special lattices for which* V_{22}^4 *is the L-polyhedron correspond to those and only those points of the C-type domain which are characterized by the condition*

$$\rho_{\overline{45}} = \rho_{\overline{12}} + \rho_{\overline{13}} + \rho_{\overline{23}}. \tag{11.8}$$

Equation (11.8) *determines a* 14*-dimensional plane in the space of coefficients of the quadratic forms.*

PROOF. Let the L-subdivision of lattices belonging to some C-type contain a skeleton of the form $\mathfrak{A}$. Without loss of generality, we will assume that this skeleton is the one-dimensional skeleton appearing in Figure 13. The vectors $l^{\overline{12}}$, $l^{\overline{13}}$, $l^{\overline{23}}$ and $l^{\overline{45}}$ are variable contiguity vectors of C-type lattices containing such a skeleton. These vectors constitute the component α.

Thus we have shown that the appearance of the body V_{22}^4 with a one-dimensional skeleton of the form $\mathfrak{A}$ in the L-subdivision of a lattice occurs only for C-types having diagrams containing a component $\kappa \sim \alpha$.

Let the diagram of some C-type domain contain a component $\kappa \sim \alpha$. We can suppose that $\kappa \equiv \alpha$ without loss of generality. Then the L-subdivision of a lattice of this C-type contains the one-dimensional skeleton represented in Figure 13. Since lattices of such a C-type do not contain any contiguity vectors from the list (11.5), according to Lemma 11.3 the four-dimensional plane (11.4) on which this skeleton is situated is laminar.

The primitive four-dimensional lattices of the plane (11.4) correspond to one of the three L-types. As follows from the results of §9.5, this L-type cannot be the type II of Voronoĭ, since it is impossible to construct a one-dimensional skeleton of six vertices from the contiguity vectors of the latter.

Turning to the contiguity vectors of type I of Voronoĭ (§3.1), i.e. to the vectors $\pm(1, 0, 0, 0), \ldots, \pm(0, 0, 0, 1), \pm(1, 1, 0, 0), \ldots, \pm(0, 0, 1, 1)$, we see that it is impossible to construct a one-dimensional skeleton of six vertices from them.

Thus only lattices of type III occur in the plane (11.4). From §9.5, the L-subdivision of lattices of this type contain the body V_{22}^4, which yields uniquely a one-dimensional skeleton of six vertices constructed from the essentially distinct contiguity vectors of such a lattice. Consequently the convex hull of the one-dimensional skeleton of Figure 13, the polyhedron $V = A_1A_2A_3B_1B_2B_3$, is the body V_{22}^4 with a basis of triangles $A_1A_2A_3$ and $B_1B_2B_3$.

Taking advantage of (3.4), we compute the squares of the lengths of the edges of the polyhedron V and on the basis of (9.1) and Lemma 9.5 we find that the body V_{22}^4 satisfies the condition that it be an L-polyhedron of a special lattice. This condition coincides with (11.8). In the usual way we verify that (11.8) defines a 14-dimensional plane in E^N ($N = 15$), and we finish the proof of the lemma.

LEMMA 11.5. *All of the C-type diagrams of Table* I *containing components* κ *equivalent to* β, *and only such figures, correspond to C-type domains whose lattices have L-subdivisions which contain the four-dimensional body* V^4_{22} *with a one-dimensional skeleton of the form* $\mathfrak{B}$.

Choose a C-type diagram such that $\kappa \equiv \beta$. *Then the four-dimensional special lattices for which* V^4_{22} *is the L-polyhedron correspond to those and only those points of the C-type domain which are characterized by the equation*

$$\rho_{\overline{12}} + \rho_{14} + \rho_{15} + \rho_{24} + \rho_{25} + \rho_{45} = 0. \tag{11.9}$$

Equation (11.9) *defines a* 14-*dimensional plane in the space of coefficients of quadratic forms.*

The proof of Lemma 11.5 is entirely analogous to the proof of Lemma 11.4, and therefore we will not carry it through. It may be reproduced without difficulty by the reader.

LEMMA 11.6. *Let the L-subdivisions of lattices of some C-type of the domain* R_1 *contain the body* V^4_{22}. *Then a decomposition of the* (*open*) *domain* Δ *of this C-type on the basis of the repartitioning complex* V^5_{22} *based on the given body* V^4_{22} *occurs if and only if* Δ *has a nonempty intersection with the* 14-*dimensional plane* Π *defined either by* (11.8) *or by* (11.9) *depending on the form of the one-dimensional skeleton of* V^4_{22}.

PROOF. The necessity is obvious. We will show sufficiency. Let Δ have a nonempty intersection with the plane Π, $\Delta \cap \Pi \neq 0$. According to Lemmas 11.4 and 11.5, the body V^4_{22} is an L-polyhedron for a lattice if and only if the lattice belongs to the set $\Delta \cap \Pi$, i.e. on this set we have a repartitioning by the complex V^4_{22} of the L-subdivision of the four-dimensional lattice.

Since the four-dimensional plane of V^4_{22} is laminar, a repartitioning in this plane implies a repartitioning of the L-subdivision of the five-dimensional lattice.

According to Lemma 10.7, the subdivision of a C-type domain of R_1 can take place only on the basis of the complex V^5_{22}. From this it follows that $\Delta \cap \Pi$ corresponds to the set of special five-dimensional lattices, where a repartitioning by V^5_{22} based on V^4_{22} occurs. The lemma is proved.

We emphasize that, due to the convexity of Δ, the "flat" 14-dimensional set $\Delta \cap \Pi$ divides Δ into two parts which are either L-type domains or the union of such domains.

We will denote the complex V^5_{22}, based on the body V^4_{22} with a one-dimensional skeleton of the form $\mathfrak{A}$, by $V^5_{22}(\mathfrak{A})$, and we will denote the complex based on the body V^4_{22} with a one-dimensional skeleton of form $\mathfrak{B}$ by $V^5_{22}(\mathfrak{B})$.

§12. The L-types of the domain R_1

1. Using Lemmas 11.4–11.6, we can clarify the situation regarding the repartitioning complexes V_{22}^5 for the set of lattices of the domain R_1 and at the same time find the decompositions of the C-types of this domain into L-types.

First of all we turn our attention to the plane (11.9) and the 14-dimensional planes which are equivalent to it. These intersect the closed domain R_1 in faces of the form H_ν (see (3.11)). From the results of §8.3 we know that the points of these faces are interior points of the C-type domains with diagrams 21, 22, 46, 47 and 74 of Table I and the C-type domains equivalent to these under the group $\mathfrak{G}$. Diagrams 21, 22, 46, 47 and 74 contain components which are equivalent to β. Therefore, according to Lemmas 11.5 and 11.6 a repartitioning takes place by means of the complex $V_{22}^5(\mathfrak{B})$ at all of the faces of the form H_ν of R_1, i.e. these faces are faces of L-type domains.

According to the results of §10, all other faces of R_1 are also faces of L-type domains. Thus we have the following theorem.

THEOREM 12.1. *The perfect domain R_1 partitions into L-type domains.*

2. We now consider all of the repartitionings using the complex $V_{22}^5(\mathfrak{B})$. Planes of the form (11.9) intersect with only 5 mutually inequivalent C-type domains with the common number 75. The diagrams of these 5 domains are given above.

Each of the diagrams 21, 22, 46, 47 and 74 contains a single component, equivalent to the component β. Moreover, they correspond to nonempty intersections of a C-type domain with a plane of the form (11.9). From this it follows that each of the five C-type domains under consideration, as we have already said, is decomposed into two parts by the complex $V_{22}^5(\mathfrak{B})$.

Many other C-type diagrams contain components which are equivalent to β. However, since these other components do not correspond to intersections of C-type domains with planes of the form (11.9), a repartitioning by the complex $V_{22}^5(\mathfrak{B})$ cannot occur.

The following 10 domains are obtained by subdividing into two parts the C-type domains with diagrams 21, 22, 46, 47 and 74: 5 correspond to intersections of the five C-type domains with R_1 (naturally the intersections are denoted by the same symbols as the domains themselves) and 5 correspond to intersections with domains which are equivalent to R_1. These last intersections are equivalent respectively to the domains (in R_1) with diagrams (a), (b), (c), 47 and 74 of Table I. Thus the C-type domains with diagrams 47 and 74 decompose into equivalent parts. We obtained this result in a different form in §8.

All additional subdivisions of C-type domains intersecting with R_1 occur via the repartitioning complex $V_{22}^5(\mathfrak{A})$. These occur only at interior points of R_1.

For the investigation of these subdivisions it is necessary to consider the 81 domains corresponding to diagrams 1–75 and (a)–(f) of Table I. The appearance in this list of domains with diagrams (a)–(c) has just been explained. The domains with diagrams (d)–(f) are portions of the C-type domains with diagrams 37, 38 and 57 (§8.4) and are obtained by subdividing these domains using faces of the perfect subdivision based on the repartitioning complex V_{23}^5 (§§10.2 and 8.4).

3. For conciseness, for the rest of this section we will call the intersection of a C-type domain with the domain R_1, represented by the diagrams of Table I, *C-type domains* and denote them by the number (or letter) of their diagram. For example, "C-type domain (a)" denotes "the intersection of the C-type domain with diagram (a) of Table I with the domain R_1".

Let Δ be a C-type domain. In order to decide whether Δ subdivides into L-type domains it is necessary, according to Lemmas 11.4 and 11.6, to find all of the components which are equivalent to α in the diagram of the domain and then check whether the 14-dimensional planes of the form (11.8) corresponding to these components intersect with Δ. In the case of a nonempty intersection the (open) domain Δ subdivides into two parts on the basis of the complex $V_{22}^5(\mathfrak{A})$. In the contrary case it does not.

We will call components of diagrams which are equivalent to α *repartitioning components,* and the 14-dimensional planes of the form (11.8) which correspond to them *separating planes.*

The following lemma tells when a separating plane intersects a C-type domain, and therefore also tells when one of these domains subdivides into L-type domains.

LEMMA 12.2. *Suppose that the diagram of the C-type domain Δ contains the repartitioning component κ. Then the separating plane Π corresponding to κ always has a nonempty intersection with Δ.*

PROOF. Without loss of generality we will assume that the component κ under consideration is the component α. Then the parameters of forms of the domain satisfy, among others, the following twenty inequalities:

$$\begin{gathered}\rho_{\overline{ij}}<\rho_{ij},\quad \rho_{k4}+\rho_{k5}+\rho_{45},\quad \rho_{\overline{k4}}+\rho_{\overline{k5}}+\rho_{45},\quad \rho_{\overline{k4}}+\rho_{k5}+\rho_{\overline{45}},\quad \rho_{k4}+\rho_{\overline{k5}}+\rho_{\overline{45}}\\ (i,\ j,\ k=1,\ 2,\ 3;\ \ i<j;\ \ k\neq i\neq j);\qquad (12.1)\\ \rho_{\overline{45}}<\rho_{45},\quad \rho_{12}+\rho_{13}+\rho_{23},\quad \rho_{\overline{12}}+\rho_{\overline{13}}+\rho_{23},\quad \rho_{\overline{12}}+\rho_{13}+\rho_{\overline{23}},\quad \rho_{12}+\rho_{\overline{13}}+\rho_{\overline{23}}.\end{gathered}$$

We remark that the parameters ρ_{45} and $\rho_{\overline{45}}$ do not appear in any inequalities defining Δ other than those listed in (12.1).

Let $f_0\in\Delta$ have the following values for its parameters:

$$\rho_{ij}=\rho_{ij}^0>0,\qquad \rho_{\overline{ij}}=\rho_{\overline{ij}}^0>0\qquad (i,\ j=1,\ldots,5;\ \ i<j).\qquad (12.2)$$

(The existence of interior points of C-type domains corresponding to the diagrams of Table I was established in §6.) Let the numbers $\rho_{\overline{45}}^0$ and ρ_{45}^0 in the collection (12.2) be replaced by the values

$$\rho_{\overline{45}}^*=\rho_{\overline{12}}^0+\rho_{\overline{13}}^0+\rho_{\overline{23}}^0,\qquad \rho_{45}^*=\max\,(\rho_{45}^0,\ \rho_{\overline{45}}^*+\varepsilon),$$

where ϵ is some arbitrary positive number.

After such a substitution the new collection of values of the parameters ρ_{ij} and

$\rho_{\overline{ij}}$ defines a point f^* of R_1 such that $f^* \in \Pi$. Substitution of these values of the parameters into inequalities (12.1) shows that they are satisfied. It follows at once that $f^* \in \Delta$. Also $\Delta \cap \Pi \neq 0$, and the lemma is proved.

The form of a repartitioning component of a diagram can naturally be distinguished as either a "free edge" or a "triangle". If a diagram is given, a free edge uniquely determines a repartitioning component. Hereafter, we will call a component with the free edge ij the *component ij*.

Two repartitioning components belonging to the same diagram may arrange themselves in one of two ways: 1) the free edges of one component may form the sides of a triangle of the other component, or 2) the situation described in 1) does not occur. In case 1) we will say that the components are connected; and in case 2), that they are disconnected.

It is easy to verify that the set of mutually disconnected components of a single diagram consists either of 2, 3, or 4 components the free edges of which emanate from a single vertex, or of three components, the free edges of which form a triangle. A set of mutually connected components may consist of only two components. It is easy to see that a third component cannot be connected simultaneously with each of a pair of components which are already connected.

When subdividing a C-type domain by a separating plane, if the component ij lies on the side of the separating plane where the inequality $\rho'_{ij} < \rho'_{km} + \rho'_{kp} + \rho'_{mp}$ (i, j, k, m, p is some permutation of 1, 2, 3, 4, 5; ρ'_{st}, as before, denotes either ρ_{st} or $\rho_{\overline{st}}$) holds, then one of the simplicial complexes of the body $V^5_{22}(\mathfrak{A})$ occurs. We will call such a complex *normal*. On the other side of the separating plane where $\rho'_{ij} > \rho'_{km} + \rho'_{kp} + \rho'_{mp}$, the other simplicial complex occurs. We will call it the *skew complex*. A simplicial complex of the body V^4_{22} with the one-dimensional skeleton of Figure 13, the simplices of which contain the triangle $A_1A_2A_3$, defines a normal complex in a five-dimensional lattice; the triangle $B_1B_2B_3$ corresponds to a skew complex.

LEMMA 12.3. *Assume that some diagram contains two connected repartitioning components α' and α''. Denote the separating planes corresponding to these components by Π' and Π''. Then there are no points of intersection of Π' and Π'' contained in the C-type domain Δ corresponding to the given diagram. Consequently Δ subdivides into only three parts. Among these parts there is none which corresponds simultaneously to skew complexes of both components κ' and κ''.*

PROOF. Let κ' have the free edge ij and κ'' have the free edge km. Then Π' and Π'' are defined by the equations

$$\rho'_{ij} = \rho'_{km} + \rho'_{kp} + \rho'_{mp}, \quad \rho'_{km} = \rho'_{ij} + \rho'_{ip} + \rho'_{jp}, \tag{12.3}$$

where i, j, k, m, p is a permutation of 1, 2, 3, 4, 5 and ρ'_{st} is either ρ_{st} or $\rho_{\overline{st}}$.

It is easy to see that the system (2.3) has a nonpositive solution only if the condition $\rho'_{kp} = \rho'_{mp} = \rho'_{ip} = \rho'_{jp} = 0$ is satisfied (which occurs at the boundary of R_1 and in fact at a face of the form F_ν (see §3)). From this it follows that Π' and Π'' do not have intersection points which are interior to R_1 and which are therefore in Δ.

From Lemma 12.2 it follows that both Π' and Π'' have interior points in Δ. Since these planes do not intersect in Δ and since Δ is convex, it follows that Π' and Π'' subdivide Δ into three parts.

The last statement of our lemma follows by substituting $>$ for the equality sign in (12.3). None of the parameters of points of Δ satisfies the resulting inequalities.

LEMMA 12.4. *Suppose the diagram of a C-type domain Δ contains the set $\kappa^{(1)}, \ldots, \kappa^{(p)}$ $(p = 2, 3, 4)$ of mutually disconnected repartitioning components. Let $\Pi^{(1)}, \ldots, \Pi^{(p)}$ be the separating planes corresponding to these components. Then $\Delta \cap \Pi^{(1)} \cap \cdots \cap \Pi^{(p)} \neq \varnothing$, and Δ is subdivided by this set of planes into 2^p parts of maximal dimensionality.*

PROOF. Consider the case where $p = 4$ or the case where $p = 3$ and the free edges of the components form a triangle. Then the set of components $\kappa^{(1)}, \ldots, \kappa^{(p)}$ determines the diagram (0-composite) and consequently the domain Δ.

Choose a point f having parameters ρ_{ij} and $\rho_{\overline{ij}}$ of the following form: the parameters corresponding to the free edges of components are equal to $3a$ $(a > 0)$; the parameters corresponding to the remaining edges are equal to a and the remaining parameters are equal to $-3a + b$ $(b > 0)$. Obviously $f \in \Delta \cap \Pi^{(1)} \cap \cdots \cap \Pi^{(p)}$, and the first assertion of our lemma is proved.

The second statement follows by choosing points f in Δ whose parameters corresponding to the free edges of the components $\kappa^{(1)}, \ldots, \kappa^{(p)}$ are exactly equal to $3a \pm \epsilon$, where $\epsilon > 0$ is chosen sufficiently small and the sign $+$ or $-$ of each of these parameters is chosen independently of the sign of all the others. The values for all other parameters are taken to be equal to those of the point f.

For the remaining cases, we denote the repartitioning components with a free black (red) edge and triangles composed of red edges by κ_{ij} $(\kappa_{\overline{ij}})$. Then the possible mutually disconnected components which have not yet been considered are the following: 1) $\kappa_{35}, \kappa_{\overline{45}}$; 2) $\kappa_{\overline{35}}, \kappa_{\overline{45}}$; 3) $\kappa_{25}, \kappa_{\overline{35}}, \kappa_{\overline{45}}$; 4) $\kappa_{\overline{25}}, \kappa_{\overline{35}}, \kappa_{\overline{45}}$. All other sets of two or three disconnected components (for three components the free edges have a common vertex) are equivalent to one of the above four.

Consider the set 4). Let f^0 be an arbitrary point of Δ whose diagram contains this set of components, and let ρ^0_{ij} and $\rho^0_{\overline{ij}}$ be the corresponding values of the parameters ρ_{ij} and $\rho_{\overline{ij}}$ of f^0.

We will replace the values $\rho^0_{\overline{25}}, \rho^0_{25}, \rho^0_{\overline{35}}, \rho^0_{35}, \rho^0_{45}$ and $\rho^0_{\overline{45}}$ by the new values

$$\begin{aligned} \rho^*_{\overline{25}} &= \rho^0_{\overline{13}} + \rho^0_{\overline{14}} + \rho^0_{\overline{34}}, & \rho^*_{25} &= \max(\rho^0_{25}, \rho^*_{\overline{25}} + \varepsilon), \\ \rho^*_{\overline{35}} &= \rho^0_{\overline{12}} + \rho^0_{\overline{14}} + \rho^0_{\overline{24}}, & \rho^*_{35} &= \max(\rho^0_{35}, \rho^*_{\overline{35}} + \varepsilon), \\ \rho^*_{\overline{45}} &= \rho^0_{\overline{12}} + \rho^0_{\overline{13}} + \rho^0_{\overline{23}}, & \rho^*_{45} &= \max(\rho^0_{45}, \rho^*_{\overline{45}} + \varepsilon), \end{aligned}$$

where $\epsilon > 0$ is arbitrary. We preserve the values of the other parameters of the previous point f^0, obtaining a new point f^* which belongs to the intersection of all three separating planes defined by the components $\kappa_{\overline{25}}, \kappa_{\overline{35}}$ and $\kappa_{\overline{45}}$. Investigation of that part of the inequalities defining Δ which is determined by the components $\kappa_{\overline{25}}, \kappa_{\overline{35}}$ and $\kappa_{\overline{45}}$ (the inequalities $\rho_{\overline{ij}} < \rho_{ij}, \rho_{km} + \rho_{kp} + \rho_{mp}, \rho_{\overline{km}} + \rho_{\overline{kp}} + \rho_{mp}, \rho_{\overline{km}} +$

$\rho_{kp} + \rho_{\overline{mp}}$ and $\rho_{km} + \rho_{\overline{kp}} + \rho_{\overline{mp}}$, where $i, j = 1, \ldots, 5, i < j$ and $ij \neq 15$) shows that $f^* \in \Delta$ and that Δ contains $2^3 = 8$ points differing from f^* only in that the parameters $\rho_{\overline{25}}, \rho_{\overline{35}}$ and $\rho_{\overline{45}}$ have the values $\rho^*_{\overline{25}} \pm \epsilon$, $\rho^*_{\overline{35}} \pm \epsilon$ and $\rho^*_{\overline{45}} \pm \epsilon$, where $\epsilon > 0$ is sufficiently small and the signs + or − are chosen in all possible ways. Thus, for the set 4) we have established the assertion of the lemma.

The proof for the remaining three sets is analogous.

From Lemmas 12.3 and 12.4 we obtain the obvious

COROLLARY. *Each C-type domain contains an L-type domain for which all of the simplicial complexes of the body $V^5_{22}(\mathfrak{A})$ occurring in the L-subdivision of the lattice are normal.*

We will say that such L-type domains are *key* for the given C-type. We will call these L-type domains *key*.

If the simplicial complex of the body $V^5_{22}(\mathfrak{A})$ is normal, then we will say that the corresponding components of the diagram are *in a normal state.* If the simplicial complex is not normal we will say that the components are *in a skew state.*

We will introduce some notation for the L-type domains of R_1. The first part of our notation designates the C-type (see §5) containing the L-type. In those cases where it is clear about which C-type we are talking, we will drop this part of our notation. The second part consists of a tabulation of the free edges of the repartitioning components of the diagram which are in a skew state, in an arbitrary order. For this we consider the diagram of the domain corresponding to the symbolic string of the domain. We will place the sign $*$ after the symbolic string of a C-type domain in the case of a key L-type domain. If the L-type domain is at the same time a C-type domain, then we will designate it by the symbolic string of the C-type domain alone, not using any additional notation.

For example, *baa0abaaaa*, 45 denotes the L-type domain contained in the C-type domain *baa0abaaaa* and whose elements satisfy the inequalities

$$\rho_{13} < \rho_{\overline{24}} + \rho_{25} + \rho_{45}, \quad \rho_{45} > \rho_{\overline{12}} + \rho_{13} + \rho_{23}.$$

In fact, the repartitioning components of the domain *baa0abaaaa* consists of the components 13 and 45. It follows from the designation of the L-type that the component 45 is in a skew state and the component 13 is in a normal state.

4. We turn now to the deduction of the L-types of the domain R_1. We need to select a subset of the set of L-types which are mutually inequivalent with respect to the group $\mathfrak{G}_1$. Each of the domains of this subset will correspond to an L-type.

As has already been mentioned, the L-type domains belonging to different C-types cannot be equivalent. For considering mutual equivalence it is therefore necessary to examine only L-types belonging to the same C-type. Two L-types will be equivalent if they interchange under a transformation of the group of automorphisms of the C-type domain to which they both belong.

We note the following obvious lemma:

LEMMA 12.5. *The group of automorphisms of the intersection of a C-type domain with the domain* R_1 *(or, as we will say in this section, the group of automorphisms of a C-type domain) is the isomorphism group of its diagram.*

The set of transformations of a diagram, as was mentioned earlier, consists of permutations of all of its vertices and recolorings at a vertex. That this set and the group $\mathfrak{G}_1$ are isomorphic is implied by Lemma 12.5.

By considering the diagrams of Table I, we discover that diagrams 1, 8, 19, 30, 32, 33, 36, 45, 48–53, 60–75 and (a)–(f) do not contain repartitioning components and consequently correspond to L-type domains. Diagrams 20, 23, 25, 27, 29, 31, 34, 35, 46, 47 and 54–59 contain a single repartitioning component, and their corresponding domains subdivide into two L-type domains which are obviously inequivalent.

The remaining 29 diagrams contain two or more repartitioning components: diagrams 9, 12, 13, 14, 15, 21, 24, 26, 28, 38, 40, 41 and 43 contain 2 components; diagrams 2, 22, 37, 39, 42 and 44 contain 3; diagrams 3, 10, 11, 16 and 17 contain 4; diagram 4 contains 5 components; diagrams 7 and 18 contain 6 components; diagram 6 contains 7 components and diagram 5 contains 10 components.

We have considered the C-type domains corresponding to these 29 diagrams. Our results are given below in Theorem 12.6 and in Table III.

Consider diagram 2. This diagram corresponds to the domain *baaaaaaaaa*. The diagram contains the components 34, 35 and 45, which are mutually disconnected. According to Lemma 12.4, this domain subdivides into the following eight L-type domains (we will drop the designation of the C-type domain in this and the following examples): $*$; 34; 35; 45; 34, 35; 34, 45; 35, 45; 34, 35, 45.

The group of diagram 2 consists of permutations of the vertices 3, 4, 5 and the permutation of the vertices 1, 2. Taking advantage of Lemma 12.5, we find that of these eight L-type domains there are only four which are mutually inequivalent: $*$; 34; 34, 35; 34, 35, 45.

Diagram 5 corresponds to the domain *bbbbbbbbbb*. The group of this diagram consists of the group of permutations of the vertices 1, 2, . . . , 5. This diagram contains the maximal number of partitioning components, which is 10.

Making use of Lemmas 12.3 and 12.4, we can find all of its L-type domains. These are the domains $*$; 12; 13; 14; 15; 23; 24; 25; 34; 35; 45, the 20 domains ij, ik, the 10 domains ij, ik, jk ($i, j, k = 1, \ldots, 5$; $i \neq j \neq k$), the 20 domains ij, ik, im ($i, j, k, m = 1, \ldots, 5$; $i \neq j \neq k \neq m$), and finally the 5 domains ij, ik, im, ip, where i, j, k, m, p is a permutation of 1, 2, 3, 4, 5. According to Lemma 12.5 we deduce that among these 66 L-type domains there are only 6 which are mutually inequivalent. These are domains 12–17 of Table III, the details of which we will give below.

The group of diagram 17 is trivial. This diagram corresponds to the domain 0*aabbbbbaa*, and its partitioning components are 13, 14, 15 and 23. Its L-type domains are $*$; 13; 14; 15; 23; 13, 14; 13, 15; 14, 15; 13, 23; 13, 14, 15. From the triviality of the group it follows that all of the 10 enumerated domains are mutually inequivalent, i.e., they define ten L-types.

Table III

The *L*-Types of the Domain in R_1

№ п/п	*C*-type	*L*-type	№ п/п	*C*-type	*L*-type	№ п/п	*C*-type	*L*-type
1	1		64	—	24	127	38	*
2	2	*	65	—	14, 24	128	—	34
3	—	34	66	—	23, 24	129	—	34, 45
4	—	34, 35	67	—	23, 25	130	39	*
5	—	34, 35, 45	68	—	23, 24, 25	131	—	12
6	3	*	69	17	*	132	—	34
7	—	13	70	—	13	133	—	45
8	—	13, 15	71	—	14	134	—	34, 45
9	4	*	72	—	15	135	40	*
10	—	13	73	—	23	136	40	34
11	—	13, 14	74	—	13, 23	137	—	45
12	5	*	75	—	13, 14	138	—	34, 45
13	—	12	76	—	13, 15	139	41	*
14	—	12, 13	77	—	14, 15	140	—	34
15	—	12, 13, 23	78	—	13, 14, 15	141	—	45
16	—	12, 13, 14	79	18	*	142	—	34, 45
17	—	12, 13, 14, 15	80	—	13	143	42	*
18	6	*	81	—	14	144	—	12
19	—	12	82	18	13, 14	145	—	34
20	—	13	83	—	13, 23	146	—	45
21	—	12, 13	84	—	14, 24	147	—	34, 35
22	—	13, 14	85	—	13, 14, 15	148	43	*
23	—	13 23	86	19		149	—	34
24	—	12, 13, 23	87	20	*	150	—	34, 45
25	—	12, 13, 14	88	—	12	151	44	*
26	—	13, 14, 15	89	21	*	152	—	12
27	—	12, 13, 14, 15	90	—	34	153	—	34
28	7	*	91	—	34, 45	154	—	34, 45
29	—	12	92	22	*	155	45	
30	—	14	93	—	12	156	46	*
31	—	12, 14	94	—	34	157	—	45
32	—	14, 24	95	—	34, 45	158	47	*
33	—	14, 34	96	23	*	159	—	45
34	—	12, 14, 24	97	—	34	160	48	
35	—	12, 24, 34	98	24	*	161	49	
36	—	12, 24, 34, 35	99	—	12	162	50	
37	8		100	—	34	163	51	
38	9	*	101	25	*	164	52	
39	—	23	102	—	34	165	53	
40	—	23, 24	103	26	*	166	54	*
41	10	*	104	—	12	167	—	45
42	—	13	105	—	34	168	55	*
43	—	13, 14	106	27	*	169	—	45
44	—	13, 23	107	—	45	170	56	*
45	11	*	108	28	*	171	—	45
46	—	14	109	28	12	172	57	*
47	—	15	110	—	45	173	—	45
48	—	14, 15	111	29	*	174	58	*
49	—	14, 24	112	—	12	175	—	45
50	12	*	113	30		176	59	*
51	—	14	114	31	*	177	—	45
52	—	14, 24	115	—	12	178	60	
53	13	*	116	32		179	61	
54	—	14	117	33		180	62	
55	13	23	118	34	*	181	63	
56	14	*	119	—	12	182	64	
57	—	14	120	35	*	183	65	
58	—	14, 24	121	35	12	184	66	
59	15	*	122	36		185	67	
60	—	15	123	37	*	186	68	
61	16	*	124	—	12	187	69	
62	—	14	125	—	34	188	70	
63	—	23	126	—	34, 45	189	71	

№ п/п	C-type	L-type	№ п/п	C-type	L-type	№ п/п	C-type	L-type
190	72		194	(a)		197	(d)	
191	73		195	(b)		198	(e)	
192	74		196	(c)		199	(f)	
193	75							

We will not give the details of our investigations of the subdivisions of all 29 C-type domains, but would like to persuade the reader, by the given example, that they are sufficiently trivial to reconstruct.

The results of our investigation are summarized in the following theorem:

THEOREM 12.6. *The domain R_1 of quadratic forms of five variables contains* 199 *L-types. The list of these L-types is contained in Table* III.

Table III is organized as follows. The column immediately following the number of the L-type includes the number of the diagram of Table I. From the diagram we can recover the symbolic string of the C-type containing the given L-type. We assume that the vertices of the diagram are enumerated in the order indicated by the "empty" diagram. The third column contains the designation of the L-type domain based on the symbolic string of the C-type domain containing it.

Consider for example the L-type domain 137 of Table III. From diagram No. 40 we find the symbolic string of the C-type to be *ba*0*aa*3*aabb*. By considering the diagram we establish that it has the two repartitioning components, 34 and 35. The designation 45 in the following column indicates that this L-type domain is defined by the inequalities

$$\rho_{34} < \rho_{\overline{12}} + \rho_{15} + \rho_{25}, \quad \rho_{\overline{45}} > \rho_{\overline{12}} + \rho_{13} + \rho_{23}. \tag{12.4}$$

Choosing a set of independent inequalities from the inequalities (12.4) and the inequalities (5.6) defining the C-type domain of diagram 40, we find that the L-type domain 137 of Table III is characterized by

$$\begin{gathered}
\rho_{15} < \rho_{\overline{15}}, \quad \rho_{25} < \rho_{\overline{25}}, \quad \rho_{\overline{35}} < \rho_{35}, \quad \rho_{34} < \rho_{\overline{34}}, \quad \rho_{\overline{12}} + \rho_{15} + \rho_{25}; \\
\rho_{23} + \rho_{25} + \rho_{35} < \rho_{14}, \quad \rho_{\overline{14}}, \quad \rho_{\overline{23}} + \rho_{25} + \rho_{\overline{35}}, \quad \rho_{23} + \rho_{\overline{25}} + \rho_{\overline{35}}; \\
\rho_{13} + \rho_{\overline{15}} + \rho_{\overline{35}} < \rho_{24}, \quad \rho_{\overline{24}}, \quad \rho_{13} + \rho_{15} + \rho_{35}, \quad \rho_{\overline{13}} + \rho_{15} + \rho_{\overline{35}}; \\
\rho_{\overline{12}} + \rho_{13} + \rho_{23} < \rho_{\overline{45}} < \rho_{45}, \quad \rho_{12} + \rho_{13} + \rho_{23}, \quad \rho_{\overline{12}} + \rho_{\overline{13}} + \rho_{23}, \quad \rho_{\overline{12}} + \rho_{13} + \rho_{\overline{23}}.
\end{gathered}$$

§13. The L-types of the primitive five-dimensional parallelohedra

1. From Theorems 10.4 and 11.1 and from the fact that the perfect domain R is an L-type domain we can deduce

THEOREM B(13.1). *In the case of quadratic forms of five variables the subdivision of the positive cone* **K** *into L-types is a subdecomposition of its subdivision into perfect domains.*

Since we have at our disposal a large amount of information about 5-dimensional parallelohedra, we will make some comments, relating to Theorem B, about the coincidences mentioned by Voronoĭ ([2], the Introduction) and about the connection between the problem of the unique subdivision of space by convex congruent polyhedra and the perfect positive quadratic forms mentioned repeatedly by Voronoĭ in [2], (the Introduction and §§104 and 118).

It appears to us that the content of Theorem B for $n = 5$ and the analogous statements for $n = 2, 3, 4$ are a reflection of a general situation, considered in Theorem 2.7, about the closure of the class of simple subdivisions of the cone $\bar{K}$ with respect to finite intersections and a reflection of the situation that the subdivision into perfect domains is the "simplest" of the simple subdivisions. We will be very happy, but surprised, if a deeper connection between these two problems is found.

In the Introduction to [2] Voronoĭ writes: "I noticed that the set of indices $\pm(p_{ik})$, $k = 1, 2, \ldots, \sigma$, is the same as the set of all representations of the minimum of a perfect quadratic form φ. The domain D corresponding to the form φ either coincides with the domain R or is a part of this domain." The coincidence stated in the quotation appears to be accidental. For example, all of the parallelohedra of R_2 have the indices

$$(0, -2, 1, 1, 1), (-2, 0, 1, 1, 1),$$
$$(-2, -2, 1, 1, 1), (0, 0, 1, 1, -1), (0, 0, 1, -1, 1), (0,0, -1, 1, 1)$$

(with respect to the coordinate system defined by the quadratic form in the form (7.1)), which are not representations of the minimum of the perfect form φ_2.

Another coincidence mentioned by Voronoĭ is: "All of the domains of the quadratic forms which I have investigated in this memoir possess the remarkable property that they are simple domains, i.e., the number of independent inequalities defining them is exactly $n(n+1)/2$." However, for $n = 5$ nonsimple L-type domains (i.e., nonsimplicial) are typical. For example, the domains of Table III numbered 1, 2, 3, 6, 7, 9, 10 and others are bounded by twenty 14-dimensional faces, and the domains numbered 4, 8, 11 and others are bounded by 21 faces. Examples of similar types of domains are mentioned by Barnes [39] and in our work [4].

2. Combining the statements of Theorems 10.9 and 12.1 and taking into account the fact that the perfect domain R is the domain of a single L-type, we formulate

THEOREM C(13.2). *In five-dimensional space there are* 221 *L-type primitive parallelohedra. The mutually inequivalent arithmetical L-type domains corresponding to the L-type primitive parallelohedra are distributed among the perfect domains* R, R_1 *and* R_2 *in the following way.* R *consists of a single L-type domain,* R_1 *contains* 199, *and* R_2 *contains* 21.

3. The geometrical construction of each of the types of primitive parallelohedra must be considered. We devote Chapter III to the solution of this problem and a

detailed description of the combinatorial metrical structure of a general five-dimensional lattice.

In concluding this chapter we remark that the construction of the L-subdivision for a rather small number of L-type five-dimensional lattices is not new. For example, the L-types of the domain of the first perfect form R (type I of Voronoĭ) have been studied for arbitrary dimension in [2], [10], [38] and [24]. Besides these, Voronoĭ [2], Ryškov [22], and Barnes and Trennery [23], [39] have described 11 additional L-types. The domains of all of these L-types belong to the perfect domain R_1.

In [2] a description of the type II of Voronoĭ for arbitrary $n \geqslant 4$ was given. In [22] descriptions of the types III and IV were given. For $n = 5$ these correspond to our L-type domains 0000*aaaaaa*, 0000*baaaaa* and 00*aa*0*aaaaa* (numbers 191, 192 and 178 of Table III). The domain of Barnes type (in [23] and [39] these are given for odd $n \geqslant 5$) for $n = 5$ coincides with the domain *aaaaaaaaaa* (number 1 of Table III). The remaining L-types of five-dimensional lattices previously known had been found in the dissertation of Trennery [23]. Using the notation of [23], they correspond to the following domains:

Δ_5 – *a*000*baaaaa*; Δ_9 – *aaa*0*ba*0*ab*0, *;
Δ_6 – 0000*baaaab*; Δ_{10} – *aaa*0*ba*0*ab*0, 23;
Δ_8 – *aa*00*aaaaaa*; Δ_{11} – *ab*00*aaaaaa*, *;
Δ_{12} – *ab*00*aaaaaa*, 45.

CHAPTER III

FORMULAS FOR THE COMBINATORIAL-METRICAL CONSTRUCTION OF THE L-SUBDIVISION OF A GENERAL FIVE-DIMENSIONAL LATTICE

§14. Laminar planes of the primitive lattices in the domain R_1

1. Laminar planes will play an important role in the investigation of the combinatorial-metrical construction of the L-subdivision of a lattice belonging to an L-type of the domain R_1. We will require more information about the organization of the C-type diagrams of Table I in order to investigate the laminar planes of lattices belonging to L-types of this domain. We will denote by $\mathfrak{G}$ or $\mathfrak{G}_\nu$ a diagram from Table I; ν is the number of the diagram in the table.

In addition to the edge or b-triangle which corresponds to it we associate with each pair of vertices i, j of an arbitrary diagram $\mathfrak{G}$ a collection of elements. These are the components of the diagram. In §§11 and 12, we presented a partial treatment of components. Namely, we considered the components α and β and those equivalent to them under the action of the group G_1. Here we will consider the general situation.

The component κ of a given diagram corresponding to the pair of vertices i, j consists of the collection of elements of the diagram for this pair and all elements which correspond to the three pairs km, kp and mp, where i, j, k, m, p is a permutation of 1, 2, 3, 4, 5. The numbers i and j of the vertices form the index of the component κ.

From the definition of component it follows that each diagram contains 10 components. We will be primarily interested in a set of components which are mutually inequivalent under the action of the group G_1. By a simple enumeration of all possibilities we find that there are 16 such components. The components $\kappa_1, \ldots, \kappa_{16}$ (Figure 15) are representatives drawn one from each equivalence class. From the form of these figures we see that the components κ_1 and κ_8 have already appeared as the components α and β.

2. We now consider the laminar planes of lattices belonging to the L-types of the perfect domain R_1. We will show that all of the laminar planes of a given L-type can be found by considering the diagram of the C-type domain which contains it.

LEMMA 14.1. *For any L-type from the domain R_1, the equation of an arbitrary laminar plane has either the form*

$$x_i - x_j = 0 \tag{14.1}$$

or the form

$$x_i + x_j = 0. \tag{14.2}$$

PROOF. Since a laminar plane must contain 15 essentially distinct contiguity vectors, Lemma 14.1 follows directly from Lemma 11.1.

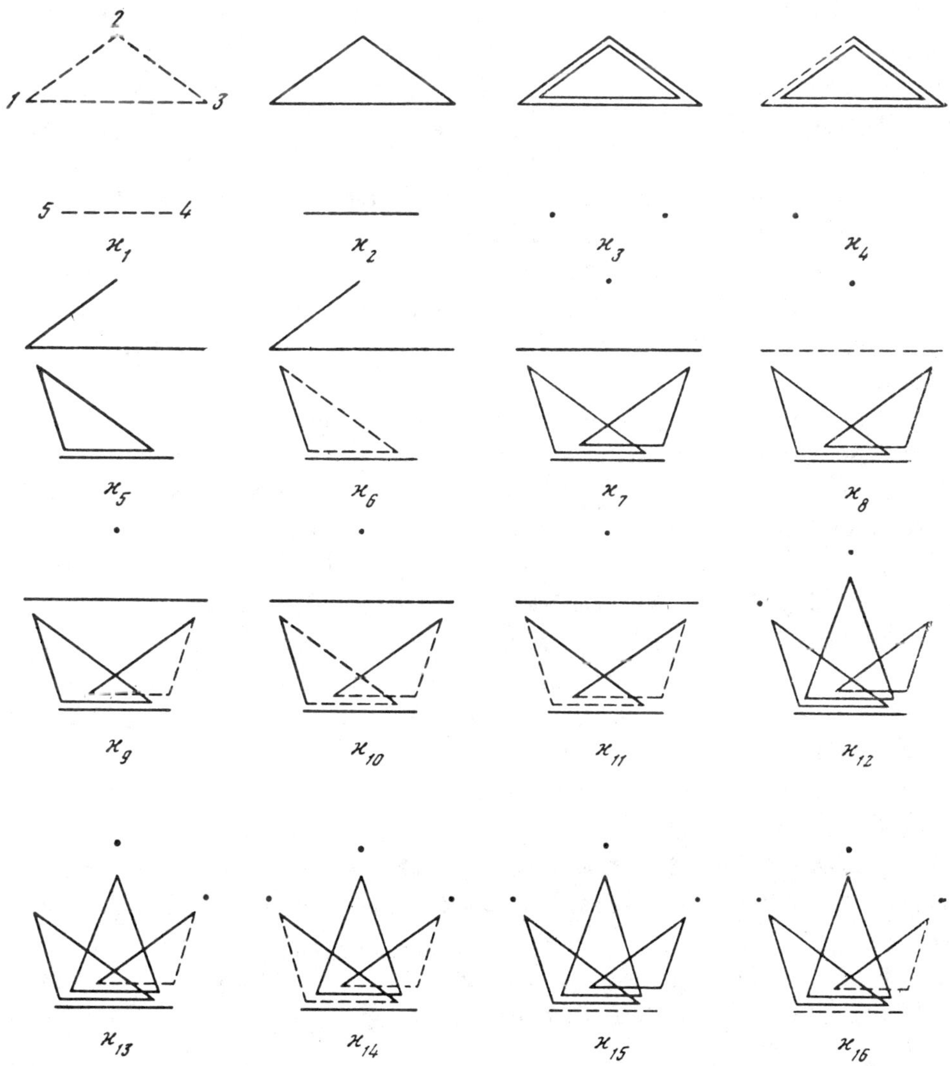

FIGURE 15

PROPOSITION 14.2. *Let* κ *be the component corresponding to the pair of vertices* i, j *of the diagram* $\mathfrak{S}$. *Then, among the laminar planes of the L-subdivision of an arbitrary lattice belonging to this C-type*: 1) *there is a plane of the form* (14.1) *and no planes of the form* (14.2) *if* $\kappa \sim \kappa_1, \kappa_2, \kappa_5, \kappa_7, \kappa_8, \kappa_{12}$ *and the edge ij of the diagram* $\mathfrak{S}$ *is black*; 2) *there is a plane of the form* (14.2) *and no planes of the form* (14.1) *if* $\kappa \sim \kappa_1, \kappa_2, \kappa_5, \kappa_7, \kappa_8, \kappa_{12}$ *but the edge ij is red*; 3) *there are planes of both forms if* $\kappa \sim \kappa_3$ *or* $\kappa \sim \kappa_4$; *and* 4) *there are planes of neither form if* $\kappa \sim \kappa_6, \kappa_9, \kappa_{10}, \kappa_{11}$ *or* $\kappa_{13}, \ldots, \kappa_{16}$.

The L-subdivisions of the lattices belonging to planes (14.1) *and* (14.2) *are related to the type* I *of Voronoĭ in the case where* $\kappa \sim \kappa_3$, *to the type* III *in the case where* $\kappa \sim \kappa_1$ *or* $\kappa \sim \kappa_8$, *and to the type* II *in all of the remaining cases.*

PROOF. In the cases where $\kappa \sim \kappa_1 = \alpha$ or $\kappa \sim \kappa_8 = \beta$ the truth of the present proposition follows from the proofs of Lemmas 11.4 and 11.5. Taking advantage of Lemma 11.3, we establish that of the two planes (14.1) and (14.2), depending on the color of the edge ij, one will be laminar and the other will not, in the case where κ is equivalent to any one of the components $\kappa_2, \kappa_5, \kappa_7$ or κ_{12}. In fact, in this case none of the vectors $l^{q,km}$, $l^{q,kp}$, $l^{q,mp}$ ($q = 0, 1, 2, 3$; i, j, k, m, p is a permutation of 1, 2, 3, 4, 5) occur among the variable contiguity vectors and, depending on the color of the edge ij, one and only one of the vectors l^{ij} and $l^{\overline{ij}}$ occurs among these variable contiguity vectors. Because of the necessity and sufficiency of the condition of Lemma 11.3 we find that the plane (14.1) is laminar if the edge ij is black but is not laminar if this edge is red, and that the plane (14.2) is not laminar if this edge is black but is laminar if it is red.

Analogously, we can prove our lemma when κ is equivalent to an arbitrary component using Lemma 11.3. For example, in the case where $\kappa \equiv \kappa_6$ ($ij = 45$) the occurrence of the vectors $l^{2,23}$ and l^{45} as contiguity vectors excludes the possibility that either of the planes $x_4 - x_5 = 0$ or $x_4 + x_5 = 0$ is laminar (see Figure 15).

The type of the four-dimensional lattice of a laminar plane can be found either by classical methods [2], [10] or by the methods given in [34]. Furthermore, in §15 we will find all of the *L*-simplices in the laminar planes for the components κ_3 and κ_4.

From Proposition 14.2 it follows that the presence of one or another laminar plane is a general property possessed by all lattices of a given *C*-type (see, for example, [18]).

Proposition 14.2 yields a simple algorithm which for a given *C*-type of the domain R_1 will quickly yield all of its laminar planes. Considering all of the diagrams of Table I, we discover the lattices belonging to the *C*-type $\mathfrak{S}_{73}$, and only they have the maximum number of laminar planes (14). These lattices coincide with those of type II of Voronoĭ.(1) Lattices with *C*-type diagrams $\mathfrak{S}_{45}, \mathfrak{S}_{60}$ and $\mathfrak{S}_{74}$ have 13 laminar planes; lattices with *C*-type diagrams with indices 19–22, 46–48, 61, 62, 66, 67, 75, (a), (b) and (c) have 12; lattices of each of the remaining *C*-types of R_1 have either 10 or 11 laminar planes.

Since each of these laminar planes corresponds, for example, to some well-defined component κ in the diagram of the *C*-type domain equivalent to one of those enumerated in Proposition 14.2, we will refer to such a laminar as *the laminar plane of component* κ.

---

(1) The coincidence of the number of laminar planes for these two types comprises a proof that the *C*-type $\mathfrak{S}_{73}$ is identical to the type II of Voronoĭ.

§15. Blocks of L-simplices.

Structural formulas for the L-subdivisions of the key L-types

1. In this and the following section we will only consider the domain R_1 and the L-types and lattices belonging to it. By a *C-type domain* we will understand the intersection a C-type domain with R_1. The points of E^S will be denoted by their radius vectors, and, as is usual, we take the point O (0, 0, 0, 0, 0) of the lattice Γ as the origin. Then the vertices of a given simplex $S \subset E^5$ will have the form $S = (\bar{a}_0, \ldots, \bar{a}_5)$.

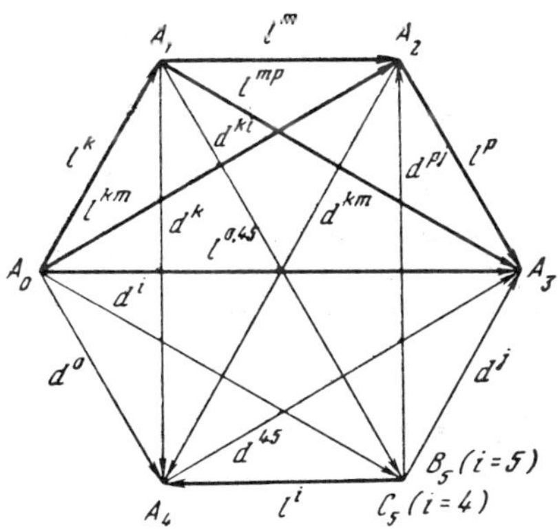

$(i, j = 4, 5\ ;\ k, m, p = 1, 2, 3)$

FIGURE 16

We introduce the following notation for the simplices of Γ (here and below we consider i, j, k, m, p to be a permutation of 1, 2, 3, 4, 5):

$$S_{ij,km} = (0,\ l^i,\ l^{ij},\ d^p,\ d^{kp},\ -d^{ij}),$$
$$S^0_{ij,km} = (0,\ l^k,\ l^{km},\ l^{0,ij},\ d^0,\ d^i),$$
$$S^1_{ij,k} = (0,\ l^m,\ l^p,\ l^{0,ij},\ d^0,\ d^i),$$
$$S^2_{ij,k} = (0,\ l^m,\ l^p,\ l^{0,ij},\ d^{ij},\ d^i),$$
$$S^3_{ij,k} = (0,\ -l^k,\ l^m,\ l^p,\ d^k,\ d^{ik}),$$
$$S^4_{ij,k} = (0,\ -l^k,\ l^m,\ l^p,\ -d^{mp},\ d^{ik}).$$

The set of 12 simplices

$$\{S^0_{ij,km},\ S^0_{ji,km},\ S^0_{ij,mk},\ S^0_{ji,mk},\ S^0_{ij,kp},\ S^0_{ji,kp},\ S^0_{ij,pk},\ S^0_{ji,pk},\ S^0_{ij,mp},\ S^0_{ji,mp},\ S^0_{ij,pm},\ S^0_{ji,pm}\} = \{ij\},$$

and all sets equivalent to it under G_1, we will call "blocks" of form $\mathfrak{L}_1$. In Figure 16 we have depicted the one-dimensional skeletons of the simplices of block {45}. The indices k, m, p of this figure are assumed to run through all permutations of 1, 2, 3.

The collection of 12 simplices

$$\{S^0_{ij,km},\ S^0_{ji,km},\ S^0_{ij,kp},\ S^0_{ji,kp},\ S^1_{ij,k},\ S^1_{ji,k},\ S^2_{ij,k},\ S^2_{ji,k},\ S^3_{ij,k},\ S^3_{ji,k},\ S^4_{ij,k},\ S^4_{ji,k}\} = [ij,\ k]$$

and all those equivalent to it under the group G_1 is called a block of form $\mathfrak{L}_2$. The one-dimensional skeletons of the simplices of block [45, 3] are depicted partly in Figure 16 and partly in Figure 17.

The four simplices

$$\{S_{ij,km},\ S_{ji,km},\ S_{ij,mk},\ S_{ji,mk}\} = (ij,\ km)$$

and all sets of four simplices equivalent to it under G_1 we will call a block of form $\mathfrak{L}_3$. The one-dimensional skeletons of the simplices of the block (ij, km) are depicted in Figure 18.

2. PROPOSITION 15.1. *Suppose that a C-type domain has among its characterizing system of inequalities the inequalities determined by the pretypes* $l^{0,ij}$, l^{km}, l^{kp}

and l^{mp} or their equivalents, i.e. the diagram of the C-type domain contains the component $\kappa \sim \kappa_3$. Then a block $\mathfrak{L}$ of the form $\mathfrak{L}_1$ is contained among the mutually non-homologous L-simplices of the lattices Γ of this C-type; $\mathfrak{L}$ can be transformed into the block $\{45\}$ by replacing κ by κ_3.

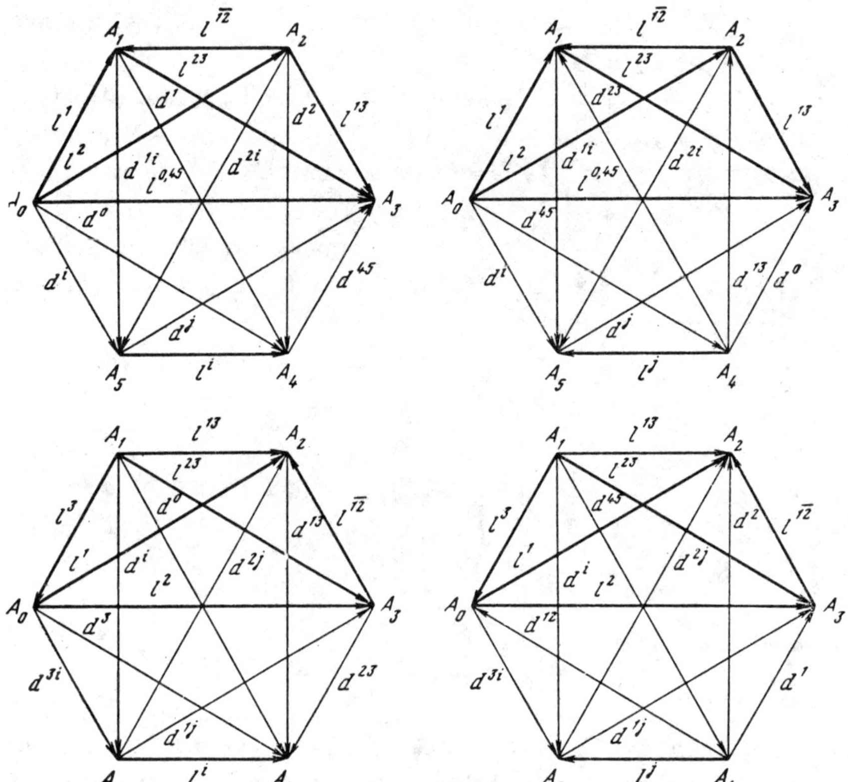

FIGURE 17

PROOF. We assume without loss of generality that $\kappa \equiv \kappa_3$. Consider one of the two laminar planes of the component κ_3, for example, the plane

$$x_4 - x_5 = 0. \tag{15.1}$$

The plane (15.1) is parallel to the following contiguity vectors:

$$l^1,\ l^2,\ l^3,\ d^0,\ d^1,\ d^2,\ d^3,\ d^{12},\ d^{13},\ d^{23},$$

$$d^{45},\ l^{12},\ l^{13},\ l^{23},\ l^{0,45}.$$

As we have already mentioned in Proposition 14.2, the L-subdivision of the plane (15.1) is related to the type I of Voronoĭ.

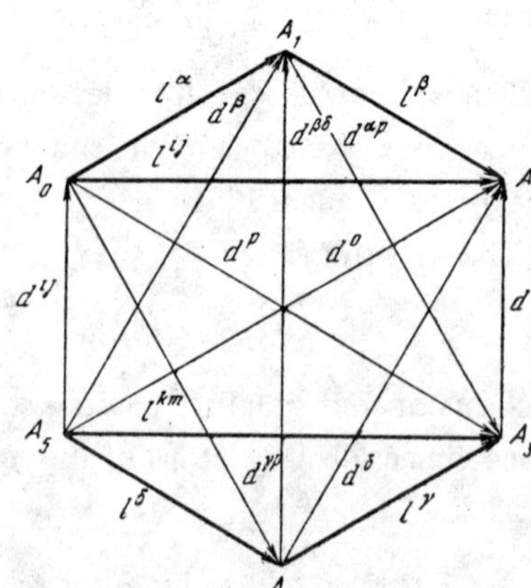

FIGURE 18

Taking advantage of [2] and the known contiguity vectors, we can find all of the mutually nonhomologous four-dimensional simplices $U_1, \ldots, U_{12}$ of the laminar plane (15.1). For convenience, we denote the vertices of each of these by the letters $A_0, \ldots, A_4$, and the edge connecting the vertices A_i and A_j ($i, j = 0, \ldots, 4; i \neq j$) by A_iA_j. In Table IV the contiguity vectors defined by the edges A_iA_j are given for each of the simplices $U_1, \ldots, U_{12}$, i.e., Table IV contains a complete description of the simplices $U_1, \ldots, U_{12}$.

TABLE IV

Simplices of block $\mathfrak{L}_1$

Simplex	A_0A_1	A_0A_2	A_0A_3	A_0A_4	A_1A_2	A_1A_3	A_1A_4	A_2A_3	A_2A_4	A_3A_4
U_1	l^1	l^{12}	$l^{0,45}$	d^0	l^2	l^{23}	d^1	l^3	d^{12}	$-d^{45}$
U_2	l^3	l^{12}	$l^{0,45}$	d^0	l^2	l^{23}	d^3	l^1	d^{23}	$-d^{45}$
U_3	l^1	l^{13}	$l^{0,45}$	d^0	l^3	l^{23}	d^1	l^2	d^{13}	$-d^{45}$
U_4	l^2	l^{13}	$l^{0,45}$	d^0	l^3	l^{23}	d^2	l^1	d^{23}	$-d^{45}$
U_5	l^2	l^{12}	$l^{0,45}$	d^0	l^1	l^{13}	d^2	l^3	d^{12}	$-d^{45}$
U_6	l^3	l^{12}	$l^{0,45}$	d^0	l^1	l^{13}	d^3	l^2	d^{13}	$-d^{45}$
U_7	l^1	l^{12}	d^0	d^3	l^2	d^1	d^{13}	d^{12}	$-d^{45}$	$-l^3$
U_8	l^2	l^{12}	d^0	d^3	l^1	d^2	d^{23}	d^{12}	$-d^{45}$	$-l^3$
U_9	l^1	l^{13}	d^0	d^2	l^3	d^1	d^{12}	d^{13}	$-d^{45}$	$-l^2$
U_{10}	l^3	l^{13}	d^0	d^2	l^1	d^3	d^{23}	d^{13}	$-d^{45}$	$-l^2$
U_{11}	l^2	l^{23}	d^0	d^1	l^3	d^2	d^{12}	d^{23}	$-d^{45}$	$-l^1$
U_{12}	l^3	l^{23}	d^0	d^1	l^2	d^3	d^{13}	d^{23}	$-d^{45}$	$-l^1$

From Table IV we find that the one-dimensional skeletons of the simplices $U_1, \ldots, U_{12}$ separate into two groups: $U_1, \ldots, U_6$ and $U_7, \ldots, U_{12}$. The simplices of the first group have one-dimensional skeletons of form 1 in Figure 19, and the simplices of the second group have one-dimensional skeletons of form 2.

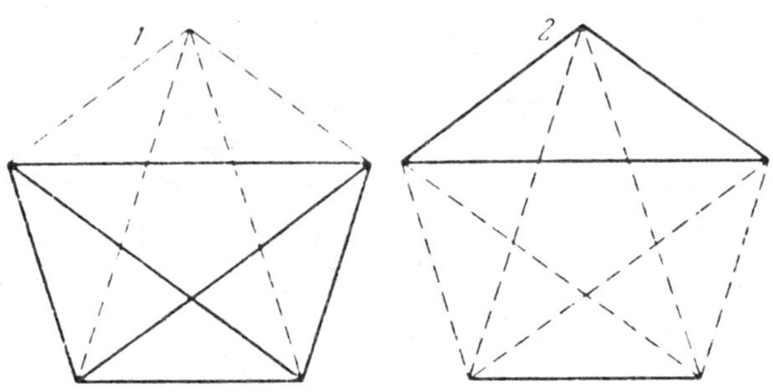

FIGURE 19

We will show that of the 12 five-dimensional L-simplices, the 6 simplices having the simplices $U_1, \ldots, U_6$ as bases have a common vertex B_5 lying in the plane

$$x_4 - x_5 = 1, \tag{15.2}$$

and that for these $\overline{A_0B_5} = (1/2, 1/2, 1/2, 1/2, -1/2)$. The remaining 6 simplices have a common vertex C_5 lying in the plane

$$x_4 - x_5 = -1, \tag{15.3}$$

and for these $\overline{A_0C_5} = (1/2, 1/2, 1/2, -1/2, 1/2)$.

Assume that for U_1 the point A_0 has coordinates (0, 0, 0, 0, 0). Then the remaining vertices of U_1 have the following coordinates:

$$A_1 = (1, 0, 0, 0, 0), \quad A_2 = (1, 1, 0, 0, 0), \quad A_3 = (1, 1, 1, 0, 0),$$
$$A_4 = (^1/_2, {}^1/_2, {}^1/_2, {}^1/_2, {}^1/_2).$$

Let S_1 be an L-simplex with base U_1 lying in the plane (5.2), and let $(x'_1, \ldots, x'_5)$ be the coordinates of its vertex B_5.

Since the vector $\overline{A_0B_5}$ must be a contiguity vector, there are three possibilities for its coordinates x'_4, x'_5:

1) $x'_4 = 0, \; x'_5 = -1$;

2) $x'_4 = 1, \; x'_5 = 0$; 3) $x'_4 = \frac{1}{2}, \; x'_5 = -\frac{1}{2}$.

Variant 1) cannot occur, since otherwise the coordinate x_5 of the vector $\overline{A_4B_5}$ would be equal to 3/2 and consequently it could not be a contiguity vector.

Taking into account the fact that the coordinates of the contiguity vector $\overline{A_3B_5}$ may not exceed 1, we find that $x'_1 \geqslant 0$, $x'_2 \geqslant 0$ and $x'_3 \geqslant 0$. Furthermore, since those cases where any of the vectors $\overline{A_iB_5}$ $(i = 0, \ldots, 4)$ with integer coordinates have fewer than two zero coordinates are prohibited, there remain only five possibilities for the coordinates of the vertex B_5:

$$1)\ (^1/_2, {}^1/_2, {}^1/_2, {}^1/_2, -{}^1/_2);\quad 2)\ (1, 0, 1, 1, 0);\quad 3)\ (1, 1, 0, 1, 0);$$
$$4)\ (1, 0, 0, 1, 0);\quad 5)\ (0, 1, 0, 1, 0).$$

We now turn to the L-simplex S'_1 with base U_1 and with vertex C_5 lying in the plane (15.3). Arguing as we did above, we find the following possibilities for the values for the coordinates of C_5:

$$1)\ (^1/_2, {}^1/_2, {}^1/_2, -{}^1/_2, {}^1/_2);\quad 2)\ (1, 0, 1, 0, 1);\quad 3)\ (1, 1, 0, 0, 1);$$
$$4)\ (1, 0, 0, 0, 1);\quad 5)\ (0, 1, 0, 0, 1).$$

Assume that one of the variants 2)–5) for the coordinates of the vertex B_5 occurs. We will show that under this assumption the vertex C_5 is determined uniquely and is equal to variant 1). In fact variants 2)–5) require that the following contiguity vectors are present in the L-subdivision:

$$\begin{aligned} &2)\ l^{1,15}, l^{\overline{24}}, l^{0,25}, l^{34}; && 4)\ l^{14}, l^{3,15}, l^{\overline{24}}; \\ &3)\ l^{24}, l^{\overline{34}}, l^{0,35}; && 5)\ l^{\overline{14}}, l^{24}, l^{3,25}, l^{1,35}. \end{aligned} \tag{15.4}$$

The contiguity vectors corresponding to variants 2)–5) of the vertex C_5 are

$$\begin{aligned} &2)\ l^{1,14}, l^{0,24}, l^{\overline{25}}, l^{35}; && 4)\ l^{3,14}, l^{15}, l^{\overline{25}}; \\ &3)\ l^{25}, l^{0,34}, l^{\overline{35}}; && 5)\ l^{\overline{15}}, l^{3,24}, l^{25}, l^{1,34}. \end{aligned} \tag{15.5}$$

Since none of the contiguity vectors (15.4) are compatible with the collection (15.5), we have proved the assertion.

Thus if any of the variants 2)–5) for the vertex B_5 occurs, then $C_5 = (1/2, 1/2, 1/2, -1/2, 1/2)$. We will now show that none of these variants are possible.

We choose a barycentric system of coordinates for the L-simplex $S_1' = C_5A_0A_1A_2A_3A_4$, writing its vertices as follows:

$$C_5=[1, 0, 0, 0, 0, 0], \quad A_0=[0, 1, 0, 0, 0, 0], \quad A_1=[0, 0, 1, 0, 0, 0],$$
$$A_2=[0, 0, 0, 1, 0, 0], \quad A_3=[0, 0, 0, 0, 1, 0], \quad A_4=[0, 0, 0, 0, 0, 1].$$

Then the variants 1)–5) of the vertex B_5 of S_1 have the following coordinates:

$$1)\ [-1, 1, 0, 0, 1, 0],\quad 2)\ [-1, 0, 1, -1, 1, 1],\quad 3)\ [-1, 0, 0, 1, 0, 1],$$
$$4)\ [-1, 0, 1, 0, 0, 1],\quad 5)\ [-1, 1, -1, 1, 0, 1]. \tag{15.6}$$

In order to decide which of the 5 points (15.6) is in fact the vertex of the L-simplex S_1, we turn, as we did in §9, to a lemma contained in [34]. According to this lemma it is necessary to find the values of the function $\Phi(x)$, given by (9.3), for each of the five points (15.6) and then compare these values. ($\Phi(x)$ is defined by (9.3) only for the case $n = 4$; in the case $n = 5$ this definition is generalized in a natural way.)

Taking advantage of the formula for the square of the length of a contiguity vector, obtained by means of (3.4), we have

$$\begin{aligned}
&f(l^i) = \sum_{\substack{\nu=1\\(\nu\neq i)}}^{5} (\rho_{i\nu} + \rho_{\overline{i\nu}}), \qquad f(d^0) = \sum_{\substack{i,j=1\\(i<j)}}^{5} \rho_{ij},\\
&f(d^i) = \sum_{\substack{\nu=1\\(\nu\neq i)}}^{5} \rho_{\overline{i\nu}} + \sum_{\substack{\mu,\tau=1\\(\mu\neq\tau\neq i,\ \mu<\tau)}}^{5} \rho_{\mu\tau},\\
&f(d^{ij}) = \rho_{ij} + \rho_{km} + \rho_{kp} + \rho_{mp} + \sum_{\substack{\nu=1\\(\nu\neq i,\ \nu\neq j)}}^{5} (\rho_{\overline{i\nu}} + \rho_{\overline{j\nu}}),\\
&f(l^{ij}) = 4\rho_{ij} + \sigma, \qquad f(l^{\overline{ij}}) = 4\rho_{\overline{ij}} + \sigma,\\
&f(l^{0,ij}) = 4(\rho_{km} + \rho_{kp} + \rho_{mp}) + \sigma,\\
&f(l^{1,ij}) = 4(\rho_{\overline{km}} + \rho_{\overline{kp}} + \rho_{mp}) + \sigma,\\
&f(l^{2,ij}) = 4(\rho_{\overline{km}} + \rho_{kp} + \rho_{\overline{mp}}) + \sigma,\\
&f(l^{3,ij}) = 4(\rho_{km} + \rho_{\overline{kp}} + \rho_{\overline{mp}}) + \sigma,
\end{aligned} \tag{15.7}$$

where

$$\sigma = \sum_{\mu=1}^{5} (\rho_{i\mu} + \rho_{j\mu} + \rho_{\overline{i\mu}} + \rho_{\overline{j\mu}}),\quad \mu\neq i,\ \mu\neq j;\ i,\ j=1,\ldots,5;\quad i\neq j,$$

Thus we find the following values for $\Phi(x)$ at the points (15.6):

1) $\Phi(-1, 1, 0, 0, 1, 0) = 2(\rho_{\overline{45}} - \rho_{12} - \rho_{13} - \rho_{23})$,
2) $\Phi(-1, 0, 1, -1, 1, 1) = (\rho_{\overline{45}} + \rho_{14} + \rho_{\overline{24}} + \rho_{34} + \rho_{\overline{23}} - \rho_{23})$,
3) $\Phi(-1, 0, 0, 1, 0, 1) = 2(\rho_{\overline{45}} + \rho_{14} + \rho_{24} + \rho_{\overline{34}})$,

4) $\Phi(-1, 0, 1, 0, 0, 1) = 2(\rho_{\overline{45}} + \rho_{14} + \rho_{\overline{24}} + \rho_{\overline{34}})$,

5) $\Phi(-1, 1, -1, 1, 0, 1) = 2(\rho_{\overline{45}} + \rho_{\overline{14}} + \rho_{24} + \rho_{34} + \rho_{\overline{12}} - \rho_{12})$.

Comparing these values and recalling that (3.4) yields a nonnegative decomposition, we see that the value $\Phi(-1, 1, 0, 0, 1, 0)$, in the case of primitive lattices, is less than any of the other four. According to [34] we know that of all the 5 points (15.6) only the point $[-1, 1, 0, 0, 1, 0]$, i.e. the point $(1/2, 1/2, 1/2, 1/2, -1/2)$, can be a vertex of the L-simplex S_1. At the same time we have shown that the two five-dimensional L-simplices having as bases the four-dimensional simplex U_1 are uniquely determined.

By considering the one-dimensional skeletons of the L-simplices S_1 and S_1' we quickly establish that $S_1 = S^0_{45,12}$ and $S_1' = S^0_{54,12}$. Since the result obtained for the four-dimensional simplex U_1 can be applied to any of the simplices $U_2, \ldots, U_6$ and the simplices of the block $\{45\}$ after an appropriate transformation from the group G_1, and since the simplices of the block $\{45\}$ are mutually nonhomologous (because they are constructed from the simplices $U_1, \ldots, U_6$ which are nonhomologous), the proof of Proposition 15.1 is finished.

We emphasize that Proposition 15.1 is not about L-types or C-types (of the domain R_1).

We remark that the structure of the L-simplices which are based on the four-dimensional simplices $U_7, \ldots, U_{12}$ depends not only on the component κ_3, but also on other elements of the C-type diagram. This follows since Proposition 15.1 exhausts all of the L-simplices with bases on the laminar plane (15.1) which are completely characterized by κ_3.

We remark further, that all of the L-simplices of the block $\{45\}$ are based on the second laminar plane $x_4 + x_5 = 0$ of the component κ_3.

3. PROPOSITION 15.2. *Let a C-type domain have among its defining inequalities the system of inequalities determined by the pretypes* $l^{0,ij}$; l^{km}, l^{kp} *and* $l^{\overline{mp}}$ *or ones equivalent to them, i.e., the diagram of the C-type domain contains the component* $\kappa \sim \kappa_4$. *Then a block* $\mathfrak{L}$ *of the form* $\mathfrak{L}_2$, *which can be transformed into* $[45, 3]$ *by transforming* κ *into* κ_4, *is contained among all of the nonhomologous L-simplices of all of the lattices of this C-type.*

PROOF. Without loss of generality we assume that $\kappa = \kappa_4$ and consider the laminar plane (15.1) of the component κ_4. The plane (15.1) is parallel to both the constant contiguity vectors $l^1, l^2, l^3, d^0, d^1, d^2, d^3, d^{12}, d^{13}, d^{23}, d^{45}$ and the variable contiguity vectors $l^{0,45}, l^{12}, l^{13}, l^{\overline{23}}$.

In the present case, as in the case of the other laminar plane of κ_4, the four-dimensional lattices contained in the laminar plane (15.1) relate to Voronoĭ's type II. Of the 12 mutually nonhomologous L-simplices of the plane (15.1), the one-dimensional skeletons of which are given in Table V, we will only be concerned with $W_1, \ldots, W_6$ (the method of constructing Table V is analogous to that used for Table IV).

By the method used in the proof of Proposition 15.1 we will investigate the five-dimensional L-simplices based on the four-dimensional simplices $W_1, \ldots, W_6$.

We consider the simplex W_1 and assume that its vertex A_0 has the coordinates $(0, 0, 0, 0, 0)$. Then the remaining vertices of this simplex have the coordinates

$$A_1=(1,\ 0,\ 0,\ 0,\ 0),\qquad A_2=(0,\ 1,\ 0,\ 0,\ 0),$$
$$A_3=(1,\ 1,\ 1,\ 0,\ 0),\qquad A_4=(^1/_2,\ ^1/_2,\ ^1/_2,\ ^1/_2,\ ^1/_2).$$

Let $B = (x'_1, \ldots, x'_5)$ be a vertex of the L-simplex D' with base W_1 which lies in the plane (15.2), and let $C = (x''_1, \ldots, x''_5)$ be a vertex of the L-simplex D'' with the same base W_1 which lies in the plane (15.3). We will show that the points B and C, independently of the C-type, are uniquely determined by the component κ_4, and furthermore that $x'_1 = \cdots = x'_4 = -x'_5 = 1/2$ and $x''_1 = \cdots = x''_3 = -x''_4 = x''_5 = 1/2$.

TABLE V

Simplices of the block $\mathfrak{L}_2$

Simplex	A_0A_1	A_0A_2	A_0A_3	A_0A_4	A_1A_2	A_1A_3	A_1A_4	A_2A_3	A_2A_4	A_3A_4
W_1	l^1	l^2	$l^{0,45}$	d^0	$-l^{\overline{12}}$	l^{23}	d^1	l^{13}	d^2	$-d^{45}$
W_2	l^1	l^2	$l^{0,45}$	d^{45}	$-l^{\overline{12}}$	l^{23}	$-d^{23}$	l^{13}	$-d^{13}$	$-d^0$
W_3	l^1	l^{13}	$l^{0,45}$	d^0	l^3	l^{23}	d^1	l^2	d^{13}	$-d^{45}$
W_4	l^1	l^{13}	$l^{0,45}$	d^{45}	l^3	l^{23}	$-d^{23}$	l^2	$-d^2$	$-d^0$
W_5	$-l^3$	l^1	l^2	d^3	l^{13}	l^{23}	d^0	$-l^{\overline{12}}$	$-d^{13}$	d^{23}
W_6	$-l^3$	l^1	l^2	$-d^{12}$	l^{13}	l^{23}	d^{45}	$-l^{\overline{12}}$	$-d^2$	$-d^1$
W_7	l^3	d^1	d^2	d^0	d^{13}	d^{23}	d^3	$l^{\overline{12}}$	l^1	l^2
W_8	l^3	d^1	d^2	$-d^{12}$	d^{13}	d^{23}	$-d^{45}$	$l^{\overline{12}}$	$-l^2$	$-l^1$
W_9	l^1	d^2	l^{13}	d^0	d^{12}	l^3	d^1	d^{45}	l^2	d^{13}
W_{10}	l^3	d^2	l^{13}	d^0	d^{23}	l^1	d^3	d^{45}	l^2	d^{13}
W_{11}	l^2	d^1	l^{23}	d^0	d^{12}	l^3	d^2	d^{45}	l^1	d^{23}
W_{12}	l^3	d^1	l^{23}	d^0	d^{13}	l^2	d^3	d^{45}	l^1	d^{23}

Taking into account that the vectors $\overline{A_iB}$ and $\overline{A_iC}$ $(i = 0, \ldots, 4)$ as contiguity vectors cannot have coordinates of modulus greater than 1 and that the coordinates of one of these vectors, if it is integral, cannot have fewer than two zeros, we find the following possibilities for the points B and C:

$$B_1=(^1/_2,\ ^1/_2,\ ^1/_2,\ ^1/_2,\ -^1/_2),\qquad C_1=(^1/_2,\ ^1/_2,\ ^1/_2,\ -^1/_2,\ ^1/_2),$$
$$B_2=(1,\ 0,\ 0,\ 1,\ 0),\qquad C_2=(1,\ 0,\ 0,\ 0,\ 1),$$
$$B_3=(0,\ 1,\ 0,\ 1,\ 0),\qquad C_3=(0,\ 1,\ 0,\ 0,\ 1),$$
$$B_4=(0,\ 0,\ 1,\ 1,\ 0),\qquad C_4=(0,\ 0,\ 1,\ 0,\ 1),$$
$$B_5=(1,\ 1,\ 0,\ 1,\ 0),\qquad C_5=(1,\ 1,\ 0,\ 0,\ 1).$$

The above variants imply that the following variable contiguity vectors will be included in the L-decomposition (among the vectors $\overline{A_iB_r}$ and $\overline{A_iC_r}$ $(i = 0, \ldots, 4;$ $r = 1, \ldots, 5))$:

B_1 — none; $\qquad$ C_1 — none;

B_2 — l^{14}, $l^{3,15}$, $l^{2,35}$; $\qquad$ C_2 — $l^{3,14}$, l^{15}, $l^{2,34}$;

B_3 — l^{24}, $l^{3,25}$, $l^{1,35}$; $\qquad$ C_3 — $l^{3,24}$, l^{25}, $l^{1,34}$;

B_4 — $l^{1,15}$, $l^{1,25}$, l^{34}, $l^{3,35}$; $\qquad$ C_4 — $l^{1,14}$, $l^{1,24}$, $l^{3,34}$, l^{35};

B_5 — l^{14}, l^{24}, $l^{\overline{34}}$, $l^{0,35}$; $\qquad$ C_5 — l^{15}, l^{25}, $l^{0,34}$, $l^{\overline{35}}$.

Comparing the collections of variable contiguity vectors, we conclude that none of the variants $B_2, \ldots, B_5$ of the vertex B are compatible with the variants $C_2, \ldots,$ C_5 of the vertex C.

We will show that none of the vertices $B_2, \ldots, B_5$ are compatible with C_1. As in the proof of Proposition 15.1, we will make use of a result from [34]. For this we choose as a basis the barycentric system of coordinates for the L-simplex $C_1A_0A_2A_3A_4$. In the barycentric coordinate system the vertices of the simplex are given by

$$C_1 = [1, 0, 0, 0, 0, 0], \quad A_0 = [0, 1, 0, 0, 0, 0], \quad A_1 = [0, 0, 1, 0, 0, 0],$$
$$A_2 = [0, 0, 0, 1, 0, 0], \quad A_3 = [0, 0, 0, 0, 1, 0], \quad A_4 = [0, 0, 0, 0, 0, 1].$$

Then the points $B_1, \ldots, B_5$ have the following barycentric coordinates:

$$B_1 = [-1, 1, 0, 0, 1, 0], \quad B_2 = [-1, 0, 1, 0, 0, 1], \quad B_3 = [-1, 0, 0, 1, 0, 1],$$
$$B_4 = [-1, 2, -1, -1, 1, 1], \quad B_5 = [-1, -1, 1, 1, 0, 1].$$

For each of these points the values of the function $\Phi(X)$ are (see the proof of Proposition 15.1):

$$\begin{aligned}
\Phi(B_1) &= 2(p_{\overline{45}} - p_{12} - p_{13} - p_{23}),\\
\Phi(B_2) &= 2(p_{\overline{45}} + p_{14} + p_{\overline{24}} + p_{\overline{34}}),\\
\Phi(B_3) &= 2(p_{\overline{45}} + p_{\overline{14}} + p_{24} + p_{\overline{34}}),\\
\Phi(B_4) &= 2(p_{\overline{45}} - p_{12} - p_{13} - p_{23} + p_{\overline{12}} + p_{\overline{13}} + p_{\overline{23}} + p_{\overline{24}} + p_{34}),\\
\Phi(B_5) &= 2(p_{\overline{45}} + p_{14} + p_{24} + p_{\overline{34}} + p_{12} - p_{\overline{12}}).
\end{aligned}$$

Recalling that the decomposition (3.4) is nonnegative and assuming that the lattice is primitive, by comparison of the values of $\Phi(X)$ we see that $\Phi(B_1) < \Phi(B_q)$, $q = 2, \ldots, 5$. According to [34] it follows that of the 5 points $B_1, \ldots, B_5$ only B_1 can be a vertex of the L-simplex D'.

Retaining the above reasoning but switching the coordinates, we show that B_1 is not compatible with any of the variants $C_2, \ldots, C_5$ for the vertex C. Thus we conclude that the L-simplices based on the four-dimensional simplex W_1 are $S^1_{45,3}$ and $S^1_{54,3}$.

As in the case of W_1, we find that the simplices $W_2, \ldots, W_6$ form bases for the following pairs of L-simplices:

$$S^2_{45,3},\ S^2_{54,3};\quad S^0_{45,13},\ S^0_{54,13};\quad S^0_{45,23},\ S^0_{54,23};\quad S^3_{45,3},\ S^3_{54,3};\quad S^4_{45,3},\ S^4_{54,3}$$

Noting that these form the mutually nonhomologous simplices of the block [45, 3], we finish the proof of Proposition 15.2.

We remark that the L-simplices based on the faces $W_7, \ldots, W_{12}$ are characterized not only by pretypes of the component κ_4 but by other pretypes.

The simplices of the block [45, 3], besides having bases on the plane (15.1), also are based on the second laminar plane $x_4 + x_5 = 0$ of the component κ_4.

4. From the one-dimensional skeletons of simplices from the block (ij, km) (Figure 18) we observe that these simplices are L-simplices of the lattice only in the case where l^{ij} and l^{km} are variable contiguity vectors of the lattice, i.e. the black edges ij and km occur among the elements of the C-type diagram to which the lattice belongs. By a transformation from G_1, coloring red one or both of the black edges ij and km, the block (ij, km) transforms into another block, its four simplices changing. We will

distinguish this new block from (ij, km) by placing a bar over the pairs of numbers denoting red edges, for example, $(\overline{ij}, km)$ or $(\overline{ij}, \overline{km})$. The simplices obtained from the simplices of the block (ij, km) by a recoloring, i.e. those with an altered sign of one or several of the variables x_i, x_j, x_k, x_m, will be denoted by putting a single bar over the index of the variable with the altered sign. The latter peculiarity in notation is explained by the following: although the block $(\overline{ij}, \ldots)$ remains unchanged independently of the fact that we have recolored the black edge ij red by changing the sign of either the variable x_i or the variable x_j, the simplex $S_{i\bar{j}}$ is not equal to the simplex $S_{\bar{i}j}$, but rather we have the equality $S_{\bar{i}j} = S_{\bar{j}i}$.

PROPOSITION 15.3. *Assume that a domain of a key L-type is such that the pretypes of index J_{ij} and J_{km} have a simple weight, i.e. its C-type diagram contains edges ij and km of a given color. Then a block $\mathfrak{L}$ of the form $\mathfrak{L}_3$ is contained among the mutually nonhomologous L-simplices of the lattice of this L-type, which can be transformed into the block (ij, km), changing ij and km into black edges.*

PROOF. Without loss of generality we assume that ij and km are the black edges 12 and 23. We will show that $S_{12,34}, S_{21,34}, S_{12,43}$ and $S_{21,43}$ are L-simplices of the lattice of the key L-type under consideration.

In order to verify that these five-dimensional simplices are L-simplices, according to [40] it is necessary and sufficient to verify that the inequalities (9)–(12) of §2 in [40] are fulfilled. These inequalities involve the squares of the lengths of the edges of the simplices.

Since the edges $A_\alpha A_\beta$ of the simplices $S_{12,34}, \ldots, S_{21,43}$ are parallel to contiguity vectors, i.e. are edges of the L-subdivision, verification of (9) is not necessary; it will be automatically fulfilled (see [40], p. 663 (transl. p. 829)). The remaining three inequalities are

$$h_{kq} + h_{kr} + h_{mq} + h_{mr} + h_{pq} + h_{pr} > h_{km} + h_{kp} + h_{mp} + h_{qr}, \tag{15.8}$$

$$\begin{aligned} 2(h_{kq} + h_{kr} + h_{ks}) + h_{mq} + h_{mr} + h_{ms} + h_{pq} + h_{pr} + h_{ps} \\ > 2(h_{km} + h_{kp}) + h_{mp} + h_{qr} + h_{qs} + h_{rs}, \end{aligned} \tag{15.9}$$

$$\begin{aligned} 2(h_{kp} + h_{kq} + h_{kr} + h_{ks}) + h_{mp} + h_{mq} + h_{mr} + h_{ms} \\ > 2h_{km} + h_{pq} + h_{pr} + h_{ps} + h_{qr} + h_{qs} + h_{rs}, \end{aligned} \tag{15.10}$$

where $h_{\alpha\beta} = A_\alpha A_\beta^2$ and k, m, p, q, r, s runs through all permutations of 0, 1, 2, 3, 4, 5.

We will assume that the forms have a nonnegative decomposition of the form (3.4), i.e. the parameters ρ_{ij} and $\rho_{\overline{ij}}$ are chosen to be nonnegative. Then, taking advantage of (15.7), we find that for each of the four simplices under consideration all of the inequalities (15.8)–(15.10), with the exception of two, are consequences of the nonnegativity of ρ_{ij} and $\rho_{\overline{ij}}$ and the inequalities which must be satisfied in order that l^{12} and l^{34} be contiguity vectors (Lemma 5.1). The two independent inequalities for the simplex $S_{12,34}$ are

$$\rho_{12} + \rho_{\overline{15}} + \rho_{25} > \rho_{34}, \qquad \rho_{34} + \rho_{\overline{35}} + \rho_{45} > \rho_{12}. \tag{15.11}$$

The independent inequalities from the set (15.8)–(15.10) for the remaining three simplices, i.e. $S_{21,34}$, $S_{12,43}$ and $S_{21,43}$, are obtained from (15.11) with the help of an appropriate reindexing. From the six inequalities obtained in this way only the following two are distinct and are not coincident with either of the inequalities (15.11):

$$\rho_{12} + \rho_{15} + \rho_{\overline{25}} > \rho_{34}, \qquad \rho_{34} + \rho_{35} + \rho_{\overline{45}} > \rho_{12}. \tag{15.12}$$

Let us assume that (15.11) or (15.12) does not hold. Let, for example,

$$\rho_{12} + \rho_{\overline{15}} + \rho_{25} \leqslant \rho_{34}. \tag{15.13}$$

From (15.3) we have

$$\rho_{\overline{15}} \leqslant \rho_{34}, \qquad \rho_{25} \leqslant \rho_{34}. \tag{15.14}$$

Comparing (15.13) and (15.14) with the trivial inequalities

$$\rho_{34} < \rho_{\overline{34}}, \quad \rho_{12} + \rho_{15} + \rho_{25}, \quad \rho_{\overline{12}} + \rho_{\overline{15}} + \rho_{25}, \quad \rho_{\overline{12}} + \rho_{15} + \rho_{\overline{25}}, \quad \rho_{12} + \rho_{\overline{15}} + \rho_{\overline{25}},$$

which form necessary and sufficient conditions that l^{34} be a contiguity vector, and noting that $\rho_{12} + \rho_{\overline{15}} + \rho_{25} \leqslant \rho_{34} < \rho_{12} + \rho_{15} + \rho_{25}$, we get

$$\begin{aligned} &\rho_{\overline{15}} \leqslant \rho_{15}, \quad \rho_{23} + \rho_{24} + \rho_{34}, \quad \rho_{\overline{23}} + \rho_{\overline{24}} + \rho_{34}, \quad \rho_{\overline{23}} + \rho_{24} + \rho_{\overline{34}}, \quad \rho_{23} + \rho_{\overline{24}} + \rho_{\overline{34}}; \\ &\rho_{25} \leqslant \rho_{\overline{25}}, \quad \rho_{13} + \rho_{14} + \rho_{34}, \quad \rho_{\overline{13}} + \rho_{\overline{14}} + \rho_{34}, \quad \rho_{\overline{13}} + \rho_{14} + \rho_{\overline{34}}, \quad \rho_{13} + \rho_{\overline{14}} + \rho_{\overline{34}}. \end{aligned}$$

Inequalities (15.15) form necessary and sufficient conditions that $l^{\overline{15}}$ and l^{25} be contiguity vectors. The contiguity vectors l^{21}, $l^{\overline{15}}$, l^{25} and l^{34} determine the repartitioning complex $V_{22}^5(A)$, which on the strength of (15.13) (see §11) occurs in a skew state.

Since we are considering a key type, in which all repartitioning complexes occur in a normal state, we have arrived at a contradiction; for this type (15.13) and analogous inequalities contradict (15.11) and (15.12), and therefore cannot hold.(2)

Since we have shown that the simplices of the block (12.34) are L-simplices in the case of a key type, we have finished the proof of Proposition 15.3.

5. Propositions 15.1–15.3 allow us, knowing the key L-types of the domain R_1, to quickly enumerate all 60 mutually nonhomologous L-simplices for lattices of these types, i.e., we are able to write down in its entirety the combinatorial-metrical structure of their L-subdivision. And, finally, we have the following structural "formula" for the L-subdivision of lattices of key L-type.

THEOREM 15.4. *The set of mutually nonhomologous L-simplices of a key L-type lattice consists of the union of blocks of the form* $\mathfrak{L}_1$, $\mathfrak{L}_2$ *and* $\mathfrak{L}_3$, *where*: 1) *blocks of form* $\mathfrak{L}_1$, *as given in Proposition* 15.1, *correspond to components* $\kappa \sim \kappa_3$ *of the diagram* $\mathfrak{S}$ *of the C-type domain containing the given key L-type*; 2) *blocks of form* $\mathfrak{L}_2$, *as given in Proposition* 15.2, *correspond to components* $\kappa \sim \kappa_4$ *of the diagram* $\mathfrak{S}$; *and* 3) *blocks of form* $\mathfrak{L}_3$, *as given in Proposition* 15.3, *correspond to pairs of edges of the diagram* $\mathfrak{S}$ *which do not have common ends.*

(2) Obviously, if (15.13) were an equality, we would have a special lattice.

PROOF. Let $\mathfrak{S}$ be the diagram of some C-type domain and s the number of pairs of edges of $\mathfrak{S}$ which do not have common ends. Then the number of components of $\mathfrak{S}$ which are equivalent to either κ_3 or κ_4 is equal to $5 - s/3$. In fact, if the C-type does not have composite pretypes, then $s = 15$. Each composite pretype corresponds either to a component $\kappa \sim \kappa_3$ or to $\kappa \sim \kappa_4$ in the diagram, and its presence implies the disappearance of an edge and therefore decreases the number s by 3. From this it follows that $t = (15 - s)/3$.

From Proposition 15.3 we know that there are $4s$ mutually nonhomologous L-simplices for a key L-type, and from Propositions 15.1 and 15.2 we know that there are $12t$; all in all there must then be $4s + 12t = 60$ such simplices. It is easy to see that the simplices from this set are mutually nonhomologous. In fact, it follows from Propositions 15.1–15.3 that the simplices of a block are nonhomologous. The fact that simplices from different blocks are nonhomologous follows from the fact that the one-dimensional skeletons of each L-simplex from a given block contain collections of edges which do not appear in the one-dimensional skeletons of the L-simplices of any other block. This finishes the proof of the theorem.

The formula for the set of mutually nonhomologous L-simplices of key L-types of Theorem 15.4 may be restated as follows:

THEOREM 15.4′. *Let $\mathfrak{S}$ be the diagram of a C-type domain which contains a given key L-type. Then each b-triangle of $\mathfrak{S}$ situated in the interior of a black triangle corresponds to a block of the form $\mathfrak{L}_1$, each b-triangle situated in the interior of a red triangle corresponds to a block of the form $\mathfrak{L}_2$, and each pair of nonintersecting edges corresponds to a block of the form $\mathfrak{L}_3$. The L-simplices obtained in this way for each block form a complete set of mutually nonhomologous L-simplices of the lattice of the given key L-type.*

We now give some examples using the formula we have just obtained.

EXAMPLE 1. *The L-type of Barnes* [23], [39]. One of his domains coincides with the C-type domain with diagram $\mathfrak{S}_1$. In $\mathfrak{S}_1$ there are no b-triangles, and all of its edges are black. By our formula the mutually nonhomologous L-simplices of lattices of this type form 15 blocks (12, 34), (12, 35), . . . , (23, 45), and they are the following:

$$S_{12,34},\ S_{21,34},\ S_{12,43},\ S_{21,43},\ \ldots,\ S_{23,45},\ S_{32,45},\ S_{23,54},\ S_{32,54}.$$

EXAMPLE 2. *The key L-type of the domain* $\underline{bbbbbbbbbb}$ *with diagram* $\mathfrak{S}_5$. The L-subdivision contains 15 blocks of the form $\mathfrak{L}_3$: $(\overline{12}, \overline{34})$, $(\overline{12}, \overline{35})$, . . . , $(\overline{23}, \overline{45})$. The mutually nonhomologous simplices are:

$$S_{\bar{1}2,\bar{3}4},\ S_{\bar{2}1,\bar{3}4},\ S_{\bar{1}2,\bar{4}3},\ S_{\bar{2}1,\bar{4}3},\ \ldots,\ S_{2\bar{3},\bar{4}5},\ S_{3\bar{2},\bar{4}5},\ S_{2\bar{3},\bar{5}4},\ S_{3\bar{2},\bar{5}4}.$$

EXAMPLE 3. *The type* II *of Voronoĭ.* The diagram of one of these domains is $\mathfrak{S}_{73}$. The four b-triangles of $\mathfrak{S}_{73}$ correspond to blocks {12}, {13}, {14} and {15}. These blocks correspond to the 48 L-simplices

$$S^0_{12,34},\ S^0_{21,34},\ S^0_{12,43},\ S^0_{21,43},\ \ldots,\ S^0_{15,34},\ S^0_{51,34},\ S^0_{15,43},\ S^0_{51,43}.$$

The remaining 12 L-simplices are contained in the blocks (23, 45), (24, 35) and (25, 34), and correspond to three pairs of black edges. These simplices are

$$S_{23,45},\ S_{32,45},\ S_{23,54},\ S_{32,54},\ \ldots,\ S_{25,34},\ S_{52,34},\ S_{25,43},\ S_{52,43}.$$

EXAMPLE 4. *The type* IV (*Ryškov* [22]). One of these domains, 0*a*0*aa*0*aaaa*, has the diagram $\mathfrak{S}_{60}$ and blocks

$$\{12\},\ \{14\},\ \{24\},\ (13,25),\ (13,45),\ (15,23),\ (15,34),\ (23,45),\ (25,34).$$

§16. The composition of the repartitioning complex $V_{22}^5(A)$

1. In this section we will establish the composition of the repartitioning complex $V_{22}^5(A)$ in both the normal and skew states. Furthermore, bearing on the combinatorial-metrical structure of the L-types of the domain R_1, the final calculations will be made for the investigation of the key L-types. The L-decomposition of an arbitrary L-type can be obtained from that of a key L-type belonging to the same C-type by substituting one or several of the complexes $V_{22}^5(A)$ in a skew state for those in a normal state.

Consider the set of lattices belonging to some L-type, the domain of which belongs to a C-type domain whose diagram contains the component κ_1. In this case, as we have established in §14, the L-subdivision of the lattices under consideration contains the laminar plane

$$x_4 + x_5 = 0. \tag{16.1}$$

The plane (16.1) is parallel (see §11) to the following 15 contiguity vectors:

$$l^1,\ l^2,\ l^3,\ l^{\overline{12}},\ l^{\overline{13}},\ l^{\overline{23}},\ l^{\overline{45}},\ d^4,\ d^5,\ d^{14},\ d^{15},\ d^{24},\ d^{25},\ d^{34},\ d^{35}.$$

As we established above (Lemma 11.4), the four-dimensional lattice belonging to the plane (16.1) is type III of Voronoĭ, and the repartitioning complex $V_{22}^5(A)$ has a base on this plane.

Depending on whether the complex $V_{22}^5(A)$ is in a normal (i.e., the inequality $\rho_{\overline{45}} < \rho_{\overline{12}} + \rho_{\overline{13}} + \rho_{\overline{23}}$ is satisfied) or a skew ($\rho_{\overline{45}} > \rho_{\overline{12}} + \rho_{\overline{13}} + \rho_{\overline{23}}$) state, the corresponding repartitioning complex V_{22}^4 in the plane (16.1) is composed of either the L-simplices T_1, T_2, T_3 or the L-simplices T'_1, T'_2, T'_3, the one-dimensional skeletons of which are given in Table VI. As in the construction of Tables IV and V, we will assume that the vertices of the simplices $T_1, \ldots, T'_3$ are denoted by the letters $A_0, \ldots, A_4$, and the edge connecting the vertices A_i and A_j ($i, j = 0, 1, \ldots, 4; i \neq j$) is denoted by A_iA_j. In Table VI we give the contiguity vectors which are parallel to the edges of the simplices. We recall (§11) that the repartitioning complex V_{22}^4 is characterized by the one-dimensional skeleton of Figure 13, where $\kappa \equiv \kappa_1$; Table VI is obtained from the simplicial complexes of the body V_{22}^4 with the one-dimensional skeleton of Figure 13 in §11.

2. LEMMA 16.1. *Choose a C-type of a lattice which contains the repartitioning complex* $V_{22}^5(A)$ *so that the complex contains the component* κ_1. *Let the vertex of the body* $V_{22}^4 \subset V_{22}^5(A)$ *corresponding to the point* A_1 *of Figure* 13 *have the coordinates* (0, 0, 0, 0, 0). *Then, independently of the C-type, the vertices of* $V_{22}^5(A)$ *not belonging to the plane of* $V_{22}^4 \subset V_{22}^5(A)$ *will have the coordinates* (1/2, 1/2, 1/2, 1/2, 1/2) *and* (1/2, 1/2, 1/2, −1/2, −1/2).

TABLE VI

Simplices of the repartitioning complex $V_{22}^5(A)$

Simplex	A_0A_1	A_0A_2	A_0A_3	A_0A_4	A_1A_2	A_1A_0	A_1A_4	A_2A_3	A_2A_4	A_3A_4
T_1	l^1	l^2	d^5	d^4	$-l^{\overline{12}}$	d^{15}	d^{14}	d^{25}	d^{24}	$-l^{\overline{45}}$
T_2	l^1	l^3	d^5	d^4	$-l^{\overline{13}}$	d^{15}	d^{14}	d^{35}	d^{34}	$-l^{\overline{45}}$
T_3	l^2	l^3	d^5	d^4	$-l^{\overline{23}}$	d^{25}	d^{24}	d^{35}	d^{34}	$-l^{\overline{45}}$
T_1'	$l^{\overline{23}}$	$l^{\overline{13}}$	d^{34}	d^{35}	$l^{\overline{12}}$	d^{24}	d^{25}	d^{14}	d^{15}	$l^{\overline{45}}$
T_2'	d^4	d^{34}	d^{24}	d^{14}	l^3	l^2	l^1	$l^{\overline{23}}$	$l^{\overline{13}}$	$l^{\overline{12}}$
T_3'	d^5	d^{35}	d^{25}	d^{15}	l^3	l^2	l^1	$l^{\overline{23}}$	$l^{\overline{13}}$	$l^{\overline{12}}$

PROOF. By assumption the point A_1 of the one-dimensional skeleton of Figure 13 has the coordinates (0, 0, 0, 0, 0). Then the other vertices of V_{22}^4 have the following coordinates:

$$A_2=(1/2,\ 1/2,\ 1/2,\ -1/2,\ 1/2),$$
$$A_3=(1/2,\ 1/2,\ 1/2,\ 1/2,\ -1/2),\ B_1=(1,\ 0,\ 0,\ 0,\ 0),$$
$$B_2=(0,\ 1,\ 0,\ 0,\ 0),\ B_3=(0,\ 0,\ 1,\ 0,\ 0).$$

Since V_{22}^4 lies in the plane $x_4+x_5=0$, one of the vertices of the polyhedron V_{22}^5, call it C_1, lies in the plane $x_4+x_5=1$. In order that $\overline{A_1C_1}$ be a contiguity vector of the lattice, only the following three variants are admissible as values of the coordinates x_4 and x_5 of the point C_1:

$$1)\ x_4=x_5=\frac{1}{2};\quad 2)\ x_4=1,\ x_5=0;\quad 3)\ x_4=0,\ x_5=1.$$

Variants 2) and 3) cannot occur, since in these cases one of the coordinates x_4 or x_5 of either the vector $\overline{A_2C_1}$ or $\overline{A_3C_1}$ would have the value 3/2 and not be a contiguity vector. This leaves only variant 1).

In order that the fractional vectors $\overline{B_1C_1}$, $\overline{B_2C_1}$ and $\overline{B_3C_1}$ be contiguity vectors, and in particular that they not have coordinates equal to 3/2, it is necessary that the coordinates x_1, x_2, x_3 of the point C_1 be equal to 1/2.

Thus we have shown that the vertex C_1 of $V_{22}^5(A)$ has the coordinates (1/2, 1/2, 1/2, 1/2, 1/2). Using analogous methods for the investigation of the position of the vertex C_2 of $V_{22}^5(A)$ in the plane $x_4+x_5=-1$, we find that its coordinates are (1/2, 1/2, 1/2, −1/2, −1/2). This finishes the proof of the lemma.

Making use of Lemma 16.1 and Table VI, we find the coordinates of the vertices of all of the L-simplices of $V_{22}^5(A)$ which are in the normal state (simplices $S_{1,1}, \ldots, S_{3,1}$ and $S_{1,2}, \ldots, S_{3,2}$) and which are in the skew state (simplices $S'_{1,1}, \ldots, S'_{3,1}$ and $S'_{1,2}, \ldots, S_{3,2}$). We tabulate these simplices, giving the radius vector of their vertices:

$$
\begin{array}{ll}
S_{1,1}=(0,\ l^1,\ l^2,\ d^5,\ d^4,\ d^0), & S_{1,2}=(0,\ l^1,\ l^2,\ d^5,\ d^4,\ d^{45}),\\
S_{2,1}=(0,\ l^1,\ l^3,\ d^5,\ d^4,\ d^0), & S_{2,2}=(0,\ l^1,\ l^3,\ d^5,\ d^4,\ d^{45}),\\
S_{3,1}=(0,\ l^2,\ l^3,\ d^5,\ d^4,\ d^0), & S_{3,2}=(0,\ l^2,\ l^3,\ d^5,\ d^4,\ d^{45}),\\
S'_{1,1}=(0,\ l^{\overline{12}},\ l^{\overline{13}},\ -d^{15},\ -d^{14},\ -d), & S'_{1,2}=(0,\ l^{\overline{12}},\ l^{\overline{13}},\ -d^{15},\ -d^{14},\ d^{23}),\\
S'_{2,1}=(0,\ l^{\overline{12}},\ l^{\overline{13}},\ -d^{15},\ l^1,\ -d^1), & S'_{2,2}=(0,\ l^{\overline{12}},\ l^{\overline{13}},\ -d^{15},\ l^1,\ d^{23}),\\
S'_{3,1}=(0,\ l^{\overline{12}},\ l^{\overline{13}},\ -d^{14},\ l^1,\ -d^1), & S'_{3,2}=(0,\ l^{\overline{12}},\ l^{\overline{13}},\ -d^{14},\ l^1,\ d^{23}).
\end{array}
$$

The one-dimensional skeletons of these simplices appear in Figure 20.

We summarize the results of this section.

PROPOSITION 16.2. *Let a C-type domain have among its defining system of inequalities those determined by the pretypes l^{ij}, $l^{\overline{km}}$, $l^{\overline{kp}}$ and $l^{\overline{mp}}$ or those determined by pretypes equivalent to them under the group G_1, i.e. the diagram $\mathfrak{S}$ of this domain contains the component $\kappa \sim \kappa_1$. Then among the mutually nonhomologous L-simplices of lattices of this C-type there are six simplices which, by a transformation $g \in G_1$ carrying the component κ into κ_1 are transformed either into the six simplices*

$$S_{1,1}, \dots, S_{3,1}, S_{1,2}, \dots, S_{3,2},$$

or into the six simplices

$$S'_{1,1}, \dots, S'_{3,1}, S'_{1,2}, \dots, S'_{3,2}$$

depending on whether the inequality obtained from the inequality $\rho_{\overline{45}} < \rho_{\overline{12}} + \rho_{\overline{13}} + \rho_{\overline{23}}$ by the transformation g holds or the inequality obtained from the inequality $\rho_{\overline{45}} > \rho_{\overline{12}} + \rho_{\overline{13}} + \rho_{\overline{23}}$ by the transformation g holds.

§17. The structure of the repartitioning complex V_{23}^5

As we have already mentioned in §3, only the C-type domains intersecting with either R_1 or R_2 contain red tetrahedrons in their diagrams. A list of these domains was given in (8.8). There the corresponding L-type domains of R_2 were given.

Distinctions between the structure of the L-subdivisions of L-type lattices of the domain τ ($\tau = 1, \dots, 21$; see §8) belonging to R_2 and lattices of one of the L-types of the C-type of R_1 which corresponds by (8.8) to τ are characterized by the repartitioning complex V_{23}^5. Thus for a description of the L-subdivision of lattices of L-type τ of R_2 we will need to know the structure of the L-type domain of R_1 which is contiguous to it, and the structure of the repartitioning complex V_{23}^5. We now consider further details of the latter.

LEMMA 17.1. *Let a C-type diagram of R_1 contain the red tetrahedron 1234 with all red edges. Then the corresponding repartitioning complex V_{23}^5 is the polyhedron $A_1A_2A_3B_1B_2B_3$ whose one-dimensional skeleton is given by the vectors*

$$\overline{B_pB_q}=l^{\overline{pq}}\ (p,\ q=1,\dots,4;\ p<q),\quad \overline{A_1A_2}=l^5,\quad A_1A_3=d^0,\quad \overline{A_2A_3}=d^5$$

(Figure 21). The simplicial complex V_{23}^5 not containing simplices of double volume (the first complex V_{23}^5) consists of four L-simplices

$$\sigma_p=A_1A_2A_3B_qB_rB_s\ (p,\ q,\ r,\ s=1,\ 2,\ 3,\ 4;\ q<r<s);$$

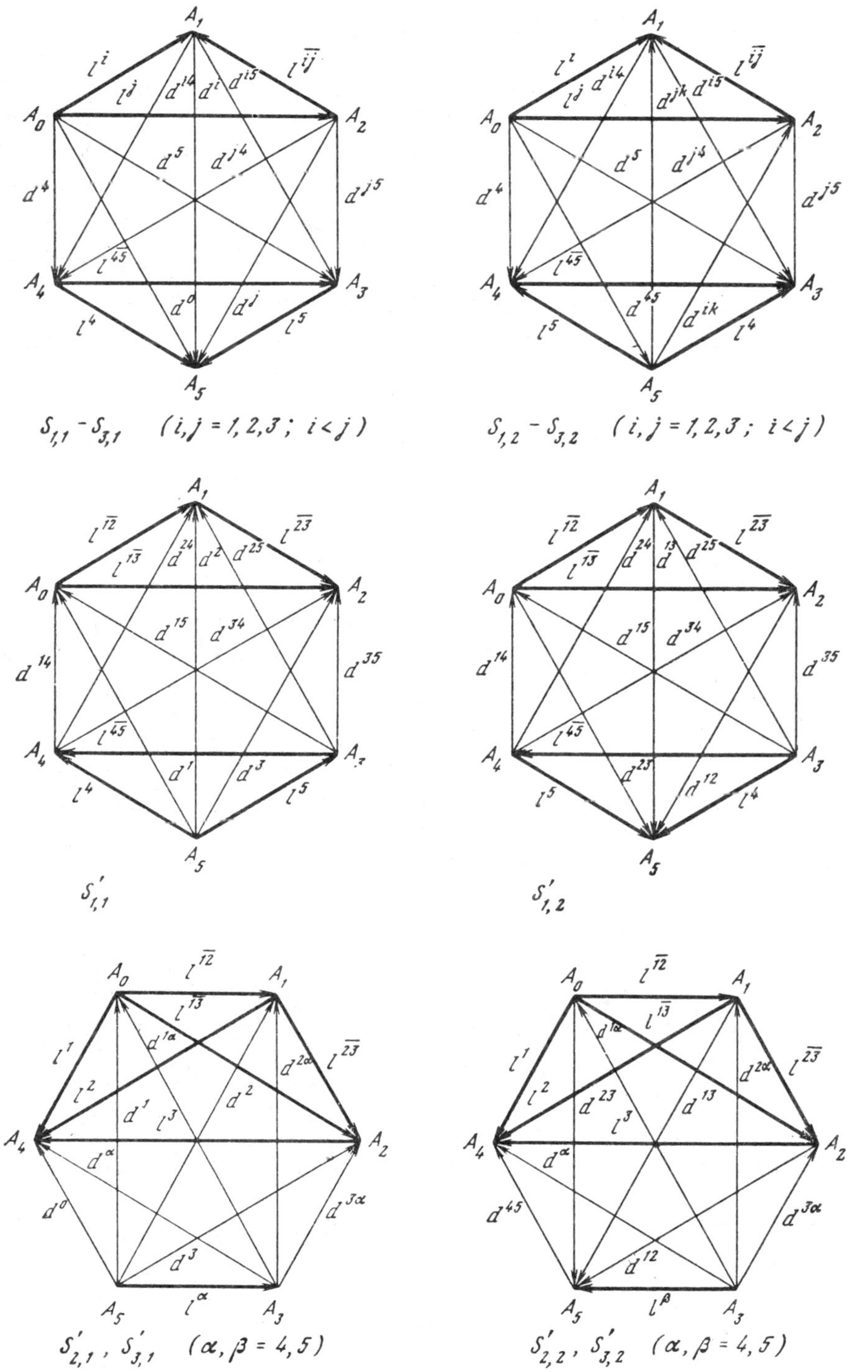

FIGURE 20

the simplicial complex containing a simplex of double volume (the second complex of V_{23}^5) consists of three L-simplices

$$B_1B_2B_3B_4A_tA_u \quad (t, u = 1, 2, 3;\ t < u),$$

of which the simplex $B_1B_2B_3B_4A_1A_2$ has double volume.

PROOF. Let Δ be an arbitrary C-type domain whose diagram $\mathfrak{S}$ contains the red tetrahedron 1234 mentioned in the lemma. We will show that an L-type appears in this C-type whose L-decomposition contains the four simplices of the first complex of V_{23}^5.

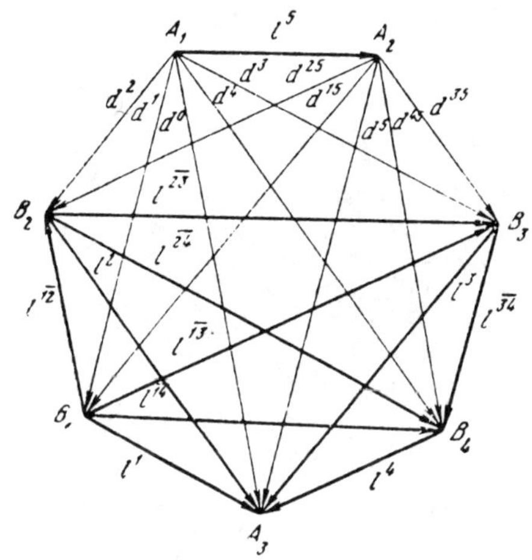

FIGURE 21

Consider the component κ corresponding to the vertices p, 5 ($p = 1, \ldots, 4$) of the diagram $\mathfrak{S}$. This component is equivalent to the component κ_1 if the pair of vertices p, 5 of $\mathfrak{S}$ corresponds to a red or black edge, and is equivalent to the component κ_4 if this pair corresponds to a b-triangle. In the first case κ corresponds to the repartitioning complex $V_{22}^5(\mathsf{A})$. The simplex σ_p is included among the six simplices of this complex in the skew state (see §16). In the second case, when $\kappa \sim \kappa_4$, the simplex σ_p is included among the simplices of a block of the form $\mathfrak{L}_2$ (Proposition 15.2).

Taking into account that the repartitioning complexes $V_{22}^5(\mathsf{A})$ corresponding to four or fewer pairs of vertices (1, 5), (2, 5), (3, 5), (4, 5) may be in a skew state (see §12), and assuming that they are in this state, we get L-types containing the L-simplices $\sigma_1, \ldots, \sigma_4$.

Consider $\sigma_1, \ldots, \sigma_4$ as L-simplices of the lattice of some L-type of R_1. For each of these the inequalities (10)–(12) of [40] are satisfied, i.e. the inequalities (15.8)–(15.10) considered above for other simplices.

For each simplex, of these inequalities, which are imposed on the squares of the lengths of the edges, we will only be interested in the following:

$$\begin{gathered} 2(A_3B_q^2 + A_3B_r^2 + A_3B_s^2) + A_1B_q^2 + A_1B_r^2 + A_1B_s^2 + A_2B_q^2 + A_2B_r^2 + A_2B_s^2 \\ > 2(A_1A_3^2 + A_2A_3^2) + A_1A_2^2 + B_qB_r^2 + B_qB_s^2 + B_rB_s^2 \\ (p,\ q,\ r,\ s = 1, \ldots, 4;\ p \neq q \neq r \neq s;\ q < r < s). \end{gathered} \tag{17.1}$$

By substituting the values of the squares of the lengths of the edges (see (15.7)) into the four inequalities (17.1) we find that they are all equivalent to the following inequality:

$$\rho_{\overline{12}} + \rho_{\overline{13}} + \rho_{\overline{14}} + \rho_{\overline{23}} + \rho_{\overline{24}} + \rho_{\overline{25}} > 0. \tag{17.2}$$

The equality corresponding to (17.2) determines a face which is equivalent to the face Q (see 8.7), i.e. a face of both of the domains R_1 and R_2' (R_2' is equivalent to R_2). But, as we know (§11), a repartitioning of the L-subdivision by the complex V_{23}^5 (and only by this complex) occurs at interior points of this face. At interior points of R_2' the inequality (17.2) is replaced by its opposite and $\sigma_1, \ldots, \sigma_4$ cease to be L-simplices.

The collection of simplices σ_p coincides with the collection of simplices of the first complex V_{23}^5 (see §10.1), from which we deduce that $\sigma_1, \ldots, \sigma_4$ form the first complex V_{23}^5.

In order to complete the proof of the lemma we must verify that the second complex V_{23}^5 coincides with that named in the lemma. We remark that the triangle $A_1A_2A_3$ appears in all of the simplices of this complex, and no other triangle appears having vertices coincident with those of V_{23}^5. From this it follows that the roles of base triangle and tetrahedron of V_{23}^5 (§9) are played by $A_1A_2A_3$ and $B_1B_2B_2B_4$. Therefore the second complex, in fact, consists of the three L-simplices which were mentioned in the statement of the lemma. That the simplex $A_1A_2B_1B_2B_3B_4$ has double volume we establish by virtue of the equality

$$V = \frac{1}{5!} \bmod \begin{vmatrix} 1 & -1 & 0 & 0 & 0 \\ 1 & 0 & -1 & 0 & 0 \\ 1 & 0 & 0 & -1 & 0 \\ -\frac{1}{2} & \frac{1}{2} & \frac{1}{2} & \frac{1}{2} & \frac{1}{2} \\ -\frac{1}{2} & \frac{1}{2} & \frac{1}{2} & \frac{1}{2} & -\frac{1}{2} \end{vmatrix} = \frac{1}{5!}$$

(we recall that the volume of the basic parallelopiped of a lattice corresponding to a quadratic form from R_1 is equal to $1/2$ because of the choice (3.3), (3.4)).

§18. The combinatorial-metrical structure of the L-subdivision of five-dimensional lattices of the domain R_1 (nonkey L-types) and of the domain R_2. Examples

1. Let some nonkey L-type Δ of R_1 belong to a C-type having the diagram $\mathfrak{S}_\nu$ ($\nu = 1, \ldots, 81$). We will call the key L-type corresponding to Δ the *key L-type of the C-type with diagram* $\mathfrak{S}_\nu$.

THEOREM 18.1. *The structure of every nonkey L-type may be found using Rule* 1.

RULE 1. Let Δ be some nonkey L-type from R_1, i.e. its number in Table III is given. Then the set $\{L\}$ of its mutually nonhomologous L-simplices is obtained from those of the set $\{L\}_\nu$ which belongs to the key L-type Δ_ν (the number ν of the diagram $\mathfrak{S}_\nu$ is given in the second column of Table III) by substituting partitioning complexes in a skew state for those in a normal state using Proposition 16.2 (the numbers ij correspond to complexes of components $\kappa \sim \kappa_1$ given in the third column of Table III).

Namely, the number ij implies that we replace the six simplices of the set $\{L\}_\nu$, obtained from the simplices

$$S_{1\bar{2},4\bar{5}}, \quad S_{1\bar{3},4\bar{5}}, \quad S_{2\bar{3},4\bar{5}}, \quad S_{1\bar{2},\bar{4}5}, \quad S_{1\bar{3},\bar{4}5}, \quad S_{2\bar{3},\bar{4}5} \tag{18.1}$$

by means of an arbitrary transformation $g \in G_1$ which transforms the component κ_1 into the component κ_{ij}, by the six L-simplices obtained from the L-simplices $S'_{1,1}, \ldots, S'_{3,1}, S'_{1,2}, \ldots, S'_{3,2}$ of (16.2) by means of the same transformation g.

PROOF. The first statement of Rule 1 is a known result from §§12 and 16 and does not require an explanation. Using Figure 20, where the one-dimensional skeletons of simplices from the list (16.2) are given, we obtain the following equalities for the first six simplices from this list:

$$\begin{aligned} S_{1,1} &= S_{1\bar{2},4\bar{5}}, & S_{1,2} &= S_{1\bar{2},\bar{4}5}, \\ S_{2,1} &= S_{1\bar{3},4\bar{5}}, & S_{2,2} &= S_{1\bar{3},\bar{4}5}, \\ S_{3,1} &= S_{2\bar{3},4\bar{5}}, & S_{3,2} &= S_{2\bar{3},\bar{4}5}. \end{aligned} \tag{18.2}$$

The second part of Rule 1 follows from these equalities, (18.2), and Proposition 16.2. This completes the proof of Theorem 18.1.

EXAMPLE 1. *The third L-type of Table* III (*L-type* 3). According to the second column of this table, $\nu = 2$ for this L-type. By Theorem 15.4 the set $\{L\}_2$ consists of the blocks

$$(\overline{12}, 34), (\overline{12}, 35), (\overline{12}, 45), (13, 24), (13, 25), (13, 45), \ldots, (23, 45).$$

In order to obtain the set $\{L\}$ for L-type 3 from $\{L\}_2$ we find the number of the component defining the repartitioning complex in the third column of Table III; this number is 34.

The component κ_{34} is obtained from κ_1 using, for example, the transformation

$$x_1 \to x_1, \quad x_2 \to x_2, \quad x_3 \to -x_5, \quad x_4 \to x_4, \quad x_5 \to -x_3. \tag{18.3}$$

Transformation (18.3) maps the simplices (18.1) onto the following simplices:

$$S_{1\bar{2},43}, \quad S_{15,43}, \quad S_{25,43}, \quad S_{1\bar{2},\bar{4}\bar{3}} = S_{1\bar{2},34}, \quad S_{15,\bar{4}\bar{3}} = S_{15,34}, \quad S_{25,\bar{4}\bar{3}} = S_{25,34}. \tag{18.4}$$

According to Rule 1, it is necessary to replace the six simplices (18.4) from the blocks $(\overline{12}, 34)$, (15, 34) and (25, 34) by the following six simplices, obtained from the simplices $S'_{1,1}, \ldots, S'_{3,1}, S'_{1,2}, \ldots, S'_{3,2}$ by means of the transformation (18.3):

$$\begin{gathered} (0, -d^{24}, -d^{14}, d^3, l^4, -l^3), \quad (0, d^{23}, d^{13}, -d^4, l^4, -l^3), \\ (0, l^1, l^2, -l^5, d^5, d^{35}), \quad (0, l^1, l^2, -l^5, d^5, d^{45}), \\ (0, l^1, l^2, -l^5, -d^{12}, d^{35}), \quad (0, l^1, l^2, -l^5, -d^{12}, d^{45}). \end{gathered}$$

EXAMPLE 2. *L-type* 14. From entry 14 of Table III we conclude that $\nu = 5$. The corresponding key L-type is described in Example 2 of §15. In the third column of Table III the numbers 12 and 13 are given; consequently the repartitioning complexes are defined by the components κ_{12} and κ_{13}. According to Rule 1 the 12 simplices

$$S_{3\bar{4},1\bar{2}}, S_{3\bar{5},1\bar{2}}, S_{4\bar{5},1\bar{2}}, S_{3\bar{4},\bar{1}2}, S_{3\bar{5},\bar{1}2}, S_{4\bar{5},\bar{1}2},$$

$$S_{2\bar{4},1\bar{3}}, S_{2\bar{5},1\bar{3}}, S_{4\bar{5},1\bar{3}}, S_{2\bar{4},\bar{1}3}, S_{2\bar{5},\bar{1}3}, S_{4\bar{5},\bar{1}3}$$

of the set $\{L\}_5$ must be replaced by the simplices

$$\begin{gathered} (0, d^3, d^4, d^5, l^1, l^2), \quad (0, d^2, d^4, d^5, l^1, l^3), \quad (0, d^{31}, d^{35}, d^{45}, l^1, l^2), \\ (0, d^{24}, d^{25}, d^{45}, l^1, l^3), \quad (0, l^3, l^4, l^5, d^1, d^0), \quad (0, l^2, l^4, l^5, d^1, d^0), \\ (0, l^3, l^4, l^5, d^2, d^0), \qquad (0, l^2, l^4, l^5, d^3, d^0), \\ (0, l^3, l^4, l^5, d^1, d^{12}), \qquad (0, l^2, l^4, l^5, d^1, d^{13}), \\ (0, l^3, l^4, l^5, d^2, d^{12}), \qquad (0, l^2, l^4, l^5, d^3, d^{13}). \end{gathered}$$

2. We refer to the L-type of the domain R_2 corresponding to diagram τ of Table II, or any domain equivalent to it under $\mathfrak{S}_2$, as the L-type τ. The L-subdivision of the L-type τ contains the repartitioning complex V_{23}^5, since the simplicial complex V_{23}^5 is (according to the terminology of §17) the second complex. Conversion of the second complex V_{23}^5 to the first translates the L-subdivision of the L-type of R_1. This L-type of R_1 is said to be contiguous with the L-type τ.

THEOREM 18.2. *The combinatorial-metrical structure of an arbitrary L-type from R_2 can be found by means of Rule* 2.

RULE 2. In order to find the combinatorial-metrical structure of the L-type τ of R_2 it is necessary:

1) to find, by means of (8.8), the number of the C-type diagram $\mathfrak{S}_\nu$ which corresponds to τ (if two C-type diagrams correspond to τ, we can choose either of them);

2) to find a red tetrahedron $ijkm$ in $\mathfrak{S}_\nu$ (choose arbitrarily if there are two or more), where i, j, k, m, p is a permutation of 1, 2, 3, 4, 5.

Then the L-type Δ_C belonging to the C-type with diagram $\mathfrak{S}_\nu$ is contiguous with the L-type τ. The repartitioning complexes $V_{22}^5(A)$ of Δ_C corresponding to the components ij, jp, kp and mp which are equivalent to κ_1 are in a skew state. Conversion of the L-subdivision of the lattice of type Δ_C into the L-subdivision of the lattice of type τ consists in replacing the four simplices of the L-subdivision of type Δ_C obtained from the simplices

$$\begin{aligned} &(0,\ d^0,\ l^p,\ d^j,\ d^k,\ d^m), \qquad (0,\ d^0,\ l^p,\ d^i,\ d^k,\ d^m),\\ &(0,\ d^0,\ l^p,\ d^i,\ d^j,\ d^m), \qquad (0,\ d^0,\ l^p,\ d^i,\ d^j,\ d^k) \end{aligned} \tag{18.5}$$

by an arbitrary transformation $g \in G_1$ which maps the tetrahedron $ijkm$ onto one with all red edges by the three simplices obtained from the simplices

$$(0,\ l^i,\ l^j,\ l^k,\ l^m,\ d^p),\quad (0,\ d^0,\ d^i,\ d^j,\ d^k,\ d^m),\quad (0,\ l^p,\ d^i,\ d^j,\ d^k,\ d^m) \tag{18.6}$$

by the same transformation g.

PROOF. From the considerations of §§8 and 9 (see Lemma 10.7 and Theorem 10.9) it follows that the contiguous L-types of the L-type τ are contained in the C-types given in (8.8), i.e. the first step of Rule 2, which finds the contiguous L-type, is justified. The second part of Rule 2 follows directly from Lemma 17.1. In fact, by replacing the red tetrahedron 1234 with red edges considered in Lemma 17.1 by the tetrahedron $ijkm$ with red edges we find that the first complex consists of the four simplices (18.5) (see Figure 21), and the second complex consists of the simplices (18.6). Thus Theorem 18.2 is proved.

EXAMPLE 3. *The L-type* 1 *of R_2*. Turning to 8.8, we see that this L-type corresponds to diagram $\mathfrak{S}_{75}$. Its red tetrahedron is the tetrahedron 1235. The transformation g, mapping all of its edges to red, we take to be

$$x_1 \to x_1,\quad x_2 \to x_2,\quad x_3 \to -x_3,\quad x_4 \to x_4,\quad x_5 \to -x_5.$$

The L-subdivision of the L-type contiguous with L-type 1 is constructed by means of the blocks

$$[14,\ 2],\ [24,\ 1],\ [34,5],\ [45,\ 3],\ (\overline{12},\ \overline{35}),\ (15,\ 32),\ (13,\ 25).$$

According to Rule 2 the following simplices, obtained from simplices (18.5) by the transformation g:

$$S^4_{14,2}=(0,\ d^{34},\ l^4,\ -d^{14},\ d^3,\ d^5),\qquad S^4_{24,1}=(0,\ d^{35};\ l^4,\ -d^{24},\ d^3,\ d^5),$$
$$S^4_{34,5}=(0,\ d^{35},\ l^4,\ -d^{14},\ -d^{24},\ d^3),\qquad S^4_{54,3}=(0,\ d^{35},\ l^4,\ -d^{14},\ -d^{24},\ d^5)$$

(each of these corresponds, respectively, to one from each of the blocks [14, 2], [24, 1], [34, 5] and [45, 3]), are replaced by the simplices

$$(0,\ l^1,\ l^2,\ -l^3,\ -l^5,\ -d^{12}),\quad (0,\ d^{35},\ -d^{14},\ -d^{24},\ d^3,\ d^5),$$
$$(0,\ l^4,\ -d^{14},\ -d^{24},\ d^3,\ d^5)$$

obtained using the transformation g from the simplices (18.6). The collection of 59 simplices obtained after this substitution yields the set of mutually nonhomologous L-simplices of lattices of L-type 1 of R_2. We mention that all of the components 14, 24, 34 and 45 are equivalent to the component κ_1, and therefore we did not consider a skew complex.

EXAMPLE 4. *L-type* 21 *of* R_2. From (8.8) we see that this type corresponds to the diagram $\mathfrak{S}_5$. In this diagram there are several red tetrahedrons. According to Rule 2 any of these tetrahedrons may be used to select the contiguous L-type. We will take tetrahedron 1234. Since all of its edges are red, the transformation g required by Theorem 18.2 is the identity transformation. The L-subdivision of the key L-type of this C-type has diagram $\mathfrak{S}_5$ and was recorded in Example 2 of §15. In order to obtain the L-subdivision of L-type 21 from this L-subdivision, first we must replace the 24 simplices appearing in the complexes $V^5_{22}(A)$ of the components 15, 25, 35 and 45 by the corresponding simplices of these same components in a skew state, and then we must replace the four simplices (18.5) appearing among these 24 simplices by the three simplices (18.6), where $i = 1, j = 2, k = 3$ and $m = 4$.

CHAPTER IV

SOLUTION OF THE PROBLEM OF THE LEAST DENSE LATTICE COVERING OF FIVE-DIMENSIONAL SPACE BY EQUAL SPHERES

§19. The result. Lemmas

1. Our results on the five-dimensional Voronoĭ-Dirichlet domains have opened up the possibility of solving a series of problems requiring all of the types of L-subdivisions of five-dimensional lattices. One of the best known of such problems is that of the value of a covering. The present chapter is devoted to this problem. Our basic result is the proof for $n = 5$ of the following theorem.

THEOREM T_n (19.1). *The minimum density of a lattice covering of E^n, $n \leqslant 5$, by equal spheres is given by the lattice Γ_n having in one of its fundamental frames the metric form*

$$\sum_{i=1}^{n} n x_i^2 - \sum_{\substack{i,\, j=1 \\ (i \neq j)}}^{n} x_i x_j$$

(*this is Voronoĭ's form of the first type*). *This minimum is equal to*

$$\theta_n^0 = \Omega_n \left[\frac{n\,(n+2)}{12\,(n+1)}\right]^{n/2} \sqrt{n+1},$$

where Ω_n denotes the volume of the n-dimensional sphere of unit radius.

We remark that for even $n \geqslant 114$ and all $n \geqslant 200$ Theorem T_n does not hold, but we will not give a proof of this fact.

Suppose that the lattice Γ is contained in E^n and that each point of this lattice is the center of a sphere of radius R. If R is chosen to be the minimum radius for which the system of spheres covers E^n, this system is called a lattice covering of E^n. The radius R is called the covering radius of Γ. The ratio $\theta_n(\Gamma)$ of the volume of the sphere of radius R to that of the fundamental region of the lattice Γ is called the density of the covering. That is,

$$\theta_n(\Gamma) = \Omega_n \frac{R^n}{\sqrt{\det (a_{ij})}}, \tag{19.1}$$

where (a_{ij}) denotes the matrix of the positive quadratic form corresponding to any of the fundamental frames of Γ.

Our task consists of finding among the set of all lattices of a given dimension that lattice yielding the least value for the density $\theta_n(\Gamma)$ or, in other words, of finding that lattice having the smallest covering radius from among those whose fundamental parallelohedra have fixed volume $V = [\det(a_{ij})]^{1/2}$.

The covering problem was solved (i.e. Theorem T_n was proved) by Kershner [42] for $n = 2$, by Bambah [43] for $n = 3$ and by Delone and Ryškov [19] for $n = 4$. References [44]–[49] are devoted to its solutions for the cases $n = 3, 4$.

That the combinatorial-metrical structure of the L-subdivision of a lattice plays a large role in the study of lattice coverings was established in the first works on coverings. This became especially clear after [50]–[52], [22] and the papers devoted to the densest lattice covering for $n = 4$.

In fact, for a given lattice Γ the radius $R = \max_t R_t$, where T_t $(t = 1, \ldots, q)$ are the radii of spheres circumscribed around the polyhedra in the L-subdivision of the lattice and

$$\theta_n(\Gamma) = \Omega_n \frac{(\max\limits_t R_t)^n}{\sqrt{\det(a_{ij})}}, \tag{19.2}$$

Therefore, in terms of the L-subdivision of a lattice, the covering problem consists in finding that lattice having a fixed volume for its fundamental parallelohedron for which the largest of the spheres circumscribed about the L-polyhedra is minimal.

The solution of the covering problem pursues the following course: First we solve the problem for the set of lattices belonging to each of the L-types of a given dimension using the known combinatorial-metrical structure of its L-subdivision. Then we choose from among the finite number of minimal coverings that one which has the smallest density, An important general result which we will need can be found in the work of Barnes and Dickson [20], who showed that each L-type has not more than one local minimum for the covering density. A simple geometrical description of this result is given in [21] and [53].

Part of the dissertation of Trenerry [23] is devoted to the determination of local minima of covering densities for the L-types of E^5 making use of the combinatorial-metrical structure of its L-subdivision. Because of computational difficulties (in spite of the use of a computer) he was unable to settle this question for the L-types of E^5 whose L-subdivisions he knew. The list of L-types he considered was far from complete.

2. The proof of Theorem T_5 would follow naturally by solving the covering problem for each of the 221 types of lattices, calculating the covering density for each of these lattices and then comparing these values. However, such an approach (which is carried out for $n = 4$ in [45], [47] and [49]) is extremely difficult for $n = 5$ (and is therefore uninteresting).

In fact, using this approach, it is necessary to overcome the following difficulties:

a) Find the radii for all of the L-simplices for each of the L-types, i.e. write out $221 \times 60 = 13260$ functions of 15 variables. Symmetry considerations, however, would help considerably in the lengthy investigation of each type.

b) Find the minima of 221 functions of 15 variables, each function being the supremum of 60 smooth functions defined on a parallelohedron.

Our solution bypasses these difficulties and is based on a series of lemmas of a different character. In this section we will introduce the underlying nature of these lemmas, giving the proof of only the first of them. The proofs of the remaining lemmas, based on more lengthy arguments, will be given in the following sections.

3. LEMMA 1 (19.2). *Suppose that the L-subdivision of the five-dimensional lattice* Γ *contains a simplex whose volume is twice that of a fundamental simplex of the lattice. Then the covering density for the lattice* Γ *is greater than* θ_5^0.

PROOF. When the fundamental parallelepiped of the lattice has unit volume the lattice Γ_n has a covering radius

$$R_n = (n+1)^{1/2n} \left[\frac{n(n+2)}{12(n+1)}\right]^{1/2},$$

and the sphere circumscribed around a proper simplex having volume $2/n!$ equal to twice that of a basic simplex of the lattice has a radius

$$r_n = 2^{1/n}(n+1)^{-1/2n}\left[\frac{n}{n+1}\right]^{1/2}.$$

Thus when $n \leqslant 5$ we have $R_n < r_n$. Because of the condition regarding the volume of a basic parallelepiped, the radius of a sphere circumscribed around any of the simplices of the lattice having double volume is not greater than r_n. From this and from (19.2) we establish the validity of the lemma.

COROLLARY. *The covering densities for the* 21 *L-type lattices contained in* R_2 *are each greater than the density* θ_5^0.

This corollary follows directly from Lemma 1 and from the fact that the L-subdivisions of all lattices corresponding to interior points in R_2 contain simplices of double volume.

4. LEMMA 2 (21.1). *When the volume of the fundamental parallelepiped of the lattice has unit volume the minimum value of the function* $(\max_t R_t)^5$, *where* R_t $(t = 1, \ldots, 6)$ *are the radii of spheres circumscribed around the L-simplices of the complex* $V_{22}^5(\mathscr{A})$, *is greater than* 0.408.

This lemma is proved in §21. From Lemma 2 it follows that the covering density $\theta_5(\Gamma)$ of Γ, whose L-subdivision contains the complex $V_{22}^5(\mathscr{A})$ in the skew state, is greater than $\Omega_5 \cdot 0.408$ and consequently greater than $\theta_5^0 = \Omega_5 \cdot 0.403$. Thus we have the

COROLLARY. *The minimum values of the covering densities for nonkey L-type lattices of* R_1 *are greater than* θ_5^0.

LEMMA 3 (24.1). *The set of all key L-type lattices of* R_1, *with the possible exception of types* 1–7, *having covering densities greater than* θ_5^0.

The proof of Lemma 3 is contained in §24 and follows from the lower bound for the density $\theta_n(\Gamma)$ given in §23 for arbitrary n of any given L-type.

LEMMA 4 (22.1). *Estimates of the minimal values of the densities $\theta_5(\Gamma)$ for the key L-types 1–7 of R_1 are as follows:*

L-type	1	2	3	4	5	6	7	
$\frac{\min_L \theta_5(\Gamma)}{\Omega_5}$	$0.423\ldots$	>0.414	>0.424	>0.427	$0.437\ldots$	>0.415	>0.413	(19.3)

This lemma is proved in §22. From (19.3) it is clear that the minimal covering densities for the lattices of the key L-types 1–7 are greater than θ_5^0. Thus the corollaries of Lemmas 1 and 2 and Lemmas 3 and 4 yield a proof of Theorem T_5.

As is clear from Lemma 4, the value of the covering density is evaluated accurately for the key L-types 1 and 5. We remark that for type 1 the value was first determined by Barnes and Trennery [39], [23].

In the proofs of Lemmas 2 and 3 an essential role is played by the uniqueness of the minimum and symmetry. We devote §20 to the consideration of these topics.

§20. Lemmas about the uniqueness of the minima and about symmetry

1. Suppose that f is the quadratic form which corresponds to some fundamental frame $\mathscr{L}$ of Γ (i.e., f is an element of the cone $\boldsymbol{K}$). Let $R(f)$ be the radius of a sphere circumscribed around the simplex constructed with the frame $\mathscr{L}$. We will consider $R(f)$ to be a function defined on $\boldsymbol{K}$.

A level surface of a function is (strictly) convex if it is the set of boundary points of a (strictly) convex body.

In [21] it is shown that the set of points f of $\boldsymbol{K}$ for which $R(f) \leqslant C$, where C is an arbitrary positive constant, is bounded and convex. Consequently the level surfaces of $R(f)$ are convex. Obviously this also applies to $R^2(f)$.

LEMMA 20.1. *The function $R^2(f)$, defined on the cone $\boldsymbol{K}$, is convex (downwards).*

PROOF. Let f' and f'' be two arbitrary points of $\boldsymbol{K}$, and let $\lambda \geqslant 0$ and $\mu \geqslant 0$ be such that $\lambda + \mu = 1$. We are required to show that

$$R^2(\lambda f' + \mu f'') \leqslant \lambda R^2(f') + \mu R^2(f''). \tag{20.1}$$

We denote by g that point of $\boldsymbol{K}$ where $f'' = C''g/C'$ if $C' = R^2(f')$ and $C'' = R^2(f'')$. Then

$$R^2(g) = R^2\left(\frac{C'}{C''} f''\right) = \frac{C'}{C''} R^2(f'') = R^2(f').$$

Further, letting $\beta = \lambda + \mu C''/C'$, we have

$$R^2(\lambda f' + \mu f'') = R^2\left(\lambda f' + \mu \frac{C''}{C'} g\right) = \beta R^2\left(\frac{\lambda}{\beta} f' + \frac{\mu C''}{\beta C'} g\right). \tag{20.2}$$

Since

$$R^2(f')=R^2(g)=C', \quad \frac{\lambda}{\beta}\geqslant 0, \quad \frac{\mu}{\beta}\frac{C''}{C'}\geqslant 0, \quad \frac{\lambda}{\beta}+\frac{\mu}{\beta}\frac{C''}{C'}=1$$

and by virtue of the convexity of the set of points for which $R^2(f)\leqslant C'$, it follows from (20.2) that

$$R^2(\lambda f'+\mu f'')\leqslant \beta C'. \tag{20.3}$$

On the other hand,

$$\lambda R^2(f')+\mu R^2(f'')=\lambda C'+\mu C''=\beta C'. \tag{20.4}$$

Comparison of (20.3) and (20.4) establishes the assertion of the lemma.

2. Let Γ_0 be an n-dimensional lattice and let $\{S_t^0\}$ $(t=1,\ldots,q)$ be a set of simplices for this lattice. We denote by $\mathscr{E}_0$ one of the frames of this lattice (fundamental or possibly a p-multiple of a fundamental frame, where p is some natural number), and by f_0 the quadratic form corresponding to this frame.

Let Γ be an arbitrary lattice and let $\mathscr{E}$ be a frame (whose multiplicity with respect to the lattice Γ is equal to that of the frame $\mathscr{E}_0$ with respect to Γ_0) with corresponding quadratic form f. Let $\{S_t\}$ denote the set of simplices the coordinates of whose vertices with respect to the frame $\mathscr{E}$ coincide with those of the simplices S_t^0 with respect to $\mathscr{E}_0$.

Let $R_t(f)$ be the radius of the sphere circumscribed around the simplex S_t. Regarding $R_t(f)$ as a function defined on the cone K, we consider the following functions:

$$\psi_1(f)=[\max_t R_t^2(f)]^n, \quad \psi_2(f)=\left[\sum_{t=1}^{q} p_t R_t^2(f)\right]^n \quad \left(\sum_{t=1}^{q} p_t=1,\ p_t\geqslant 0\right). \tag{20.5}$$

LEMMA 20.2. *The level surfaces of the functions $\psi_1(f)$ and $\psi_2(f)$ on the cone K are convex.*

PROOF. First we will show that each of the functions $R_t^2(f)$ is convex. Let $\mathfrak{g}$ be a linear transformation such that $\mathfrak{g}f=f_t$, where f_t is the positive quadratic form corresponding to one of the frames of the simplex S_t, i.e. one of the frames of Γ. Then $R_t^2(f)=R^2(f_t)=R^2(\mathfrak{g}f)$. According to Lemma 20.1 the function $R^2(f_t)$, as a function of f_t, is convex, i.e. for arbitrary points $f_t'=\mathfrak{g}f'$ and $f_t''=\mathfrak{g}f''$ of K and for arbitrary $\lambda\geqslant 0$ and $\mu\geqslant 0$, $\lambda+\mu=1$, the following inequality holds:

$$R^2(\lambda f_t'+\mu f_t'')\leqslant \lambda R^2(f_t')+\mu R^2(f_t''). \tag{20.6}$$

Since

$$R^2(\lambda f_t'+\mu f_t'')=R^2(\lambda\mathfrak{g}f'+\mu\mathfrak{g}f'')=R^2[\mathfrak{g}(\lambda f'+\mu f'')]=R_t^2(\lambda f'+\mu f''),$$

we have, using (20.6),

$$R_t^2(\lambda f'+\mu f'')\leqslant \lambda R_t^2(f')+\mu R_t^2(f''),$$

and this establishes the convexity of $R_t^2(f)$.

Let t_1 be that number for which

$$\max_t R_t(\lambda f' + \mu f'') = R_{t_1}(\lambda f' + \mu f''). \tag{20.7}$$

Then, since $R_{t_1}^2(f)$ is convex and since the function $y = x^n$ is monotonely increasing and convex, we have (see [54])

$$[R_{t_1}^2(\lambda f' + \mu f'')]^n \leqslant \lambda [R_{t_1}^2(f')]^n + \mu [R_{t_1}^2(f'')]^n. \tag{20.8}$$

Since

$$R_{t_1}^2(f) \leqslant \max_t R_t^2(f),$$

we have from (20.7) and (20.8) that

$$\psi_1(\lambda f' + \mu f'') \leqslant \lambda \psi_1(f') + \mu \psi_1(f''),$$

and the convexity of $\psi_1(f)$ is established. The function $\psi_2(f)$ is convex since $\Sigma_1^q p_t R_t^2(f)$, being a positive sum of convex functions, is convex, and since $y = x^n$ is strictly monotone and convex.

From the convexity of $\psi_1(f)$ and $\psi_2(f)$ it follows that their level surfaces are convex.

3. We will make use of the following functions, defined on K, to estimate the covering densities of lattices of a given L-type:

$$\begin{gathered}\varphi_1(f) = \Omega_n \left(\frac{\psi_1(f)}{\det f}\right)^{1/n} = \Omega_n \frac{\left[\max_t R_t^2(f)\right]^{n/2}}{(\det f)^{1/2}}, \\ \varphi_2(f) = \Omega_n \left(\frac{\psi_2(f)}{\det f}\right)^{1/n} = \Omega_n \frac{\left[\sum_{t=1}^q p_t R_t^2(f)\right]^{n/2}}{(\det f)^{1/2}} \\ \left(\sum_{t=1}^q p_t = 1,\ p_t \geqslant 0\right).\end{gathered} \tag{20.9}$$

We remark that use of the function $\varphi_2(f)$ as an estimator of the covering density is a generalization of a method proposed by B. N. Delone and used in [51] to investigate the covering densities of Voronoï's type I lattices.

We cite below some properties of the functions (20.9) which we will need. Since $\varphi_i(f) \equiv \varphi_i(kf)$, where k is an arbitrary positive number, in the investigation of these functions on the cone K, as is often done, we will refer to the ray $kf \subset K$ as a point.

LEMMA 20.3. *The functions $\varphi_1(f)$ and $\varphi_2(f)$ have unique minima.*

PROOF. The unbounded closed body D defined by the equation $\det f \geqslant d_0$ ($d_0 = \text{const} > 0$) is strictly convex according to a lemma of Minkowski [29], i.e. the level surfaces of $\det f$ are strictly convex. According to Lemma 20.2 the level surfaces of $\psi_1(f)$ and $\psi_2(f)$ are convex. Since the body D does not contain the vertex O of K while the closures of the bodies $\psi_1(f) \leqslant C_1$ and $\psi_2(f) \leqslant C_2$ do contain O, the body D has unique contact points with the bodies $\psi_1 \leqslant C_1$ and $\psi_2(f) \leqslant C_2$ for arbitrary d_0 and suitable choices of C_1 and C_2. Obviously these points are unique minima of the functions $\psi_1(f)/\det f$ and $\psi_2(f)/\det f$. The uniqueness of the minima of $\varphi_1(f)$ and $\varphi_2(f)$ follows.

LEMMA 20.4. *Let H be the group of affine automorphisms of the set of simplices $\{S_t\}$. Then the point f_0 at which the minimum of $\varphi_i(f)$ $(i = 1, 2)$ is achieved is invariant with respect to the group $\mathfrak{H}$ of transformations of K induced by H, i.e. to the point f_0 corresponds that lattice whose group of metric automorphisms is H.*

PROOF. Let $\mathfrak{h} \in \mathfrak{H}$. Then $\mathfrak{h} f_0 = f_1$. Since $\varphi_i(f_1) = \varphi_i(f_0)$, it follows from Lemma 20.3 that $f_1 = f_0$, which proves Lemma 20.4.

4. As we mentioned earlier, the functions $\varphi_i(f)$ are single valued on the ray kf $(0 < k < +\infty)$, and we can therefore talk about the level cones of these functions.

LEMMA 20.5. *The level cones of $\varphi_1(f)$ and $\varphi_2(f)$ are strictly convex.*

PROOF. Obviously the level cones of $\varphi_i(f)$ $(i = 1, 2)$ are the same as those of the functions $\widetilde{\varphi}_i(f) = [\varphi_i(f)/\Omega_n]^2$. Therefore a proof of the strict convexity of the latter will serve as a proof of our lemma. Let F_C be a level cone of the function $\widetilde{\varphi}_i(f)$, i.e. F_C is the set of points f where

$$\widetilde{\varphi}_i(f) = \left[\frac{\varphi_i(f)}{\Omega_n}\right]^2 = \frac{\psi_i(f)}{\det f} = C.$$

Suppose that f and g are arbitrary points of F_C not lying on the same ray. Let g_1 be such that $g_1 = k_1 f$ and $\det g_1 = \det f$. Choose arbitrary $\lambda > 0$ and $\mu > 0$, $\lambda + \mu = 1$. Then the numbers $k_2 = (\mu + \lambda k_1)^{-1}$, $\lambda_1 = \lambda k_1 k_2$ and $\mu_1 = \mu k_2$ satisfy $\mu_1 + \lambda_1 = 1$ and $\lambda_1 f + \mu_1 g_1 = k_1 k_2(\lambda f + \mu g)$. From the strict convexity of the equidiscriminant surface and from the convexity of $\psi_i(f)$ it follows that

$$\det f = \det g_1 = \lambda_1 \det f + \mu_1 \det g_1 < \det(\lambda_1 f + \mu_1 g_1),$$

and

$$\lambda_1 \psi_i(f) + \mu_1 \psi_i(g_1) \geqslant \psi_i(\lambda_1 f + \mu_1 g_1).$$

Furthermore,

$$\lambda \widetilde{\varphi}_i(f) + \mu \widetilde{\varphi}_i(g) = C = \lambda_1 \widetilde{\varphi}_i(f) + \mu_1 \widetilde{\varphi}_i(g_1) = \frac{\lambda_1 \psi_i(f) + \mu_1 \psi_i(g_1)}{\det f}$$

$$> \frac{\psi_i(\lambda_1 f + \mu_1 g_1)}{\det(\lambda_1 f + \mu_1 g_1)} = \widetilde{\varphi}_i(\lambda_1 f + \mu_1 g_1) = \widetilde{\varphi}_i(\lambda f + \mu g),$$

and this yields the proof of Lemma 20.5.

Lemma 20.5 can be used for developing "machine" methods for finding the minima of the functions $\varphi_i(f)$; that is, these minima can be found to within a defined accuracy using a computer.

5. As in subsection 2, we assume $\{S_t\}$ to be a set of simplices of a star in the L-subdivision of the lattice Γ and in fact a star with center at the point 0 $(0, 0, \ldots, 0)$. We denote by δ the set of lattices Γ the stars of whose L-subdivisions contain the set $\{S_t\}$. Since according to our condition the coordinates of the vertices of the simplices for the various lattices must be identical (that is, the lattices of the set δ considered with their respective frames), there will be a domain Δ_δ of K which corresponds to δ.

Obviously Δ_δ is the union of L-type domains and is convex. In fact, if $\{S_t\}$ is the totality of L-simplices of some lattice Γ_0 represented by the point f_0 in Δ_δ, then an arbitrary point f belonging to the L-type domain to which f_0 belongs has a lattice for which $\{S_t\}$ is the collection of L-simplices. The convexity of Δ_δ follows from the linearity of the inequalities, which requires that a sphere circumscribed around a given simplex of the lattice be empty.

We remark that Δ_δ is not only the union of L-type domains, but the union of a finite number of L-type domains, and on the strength of its convexity it is bounded by a finite number of linear inequalities. However, this fact will not be needed in the sequel, and so we will leave the proof to the reader.

Let $\mathfrak{G}_\delta$ be the group of integral unimodular (if the frame $\mathscr{L}$ is fundamental, or of an appropriate type if the frame has multiplicity p) automorphisms of the set Δ_δ.

LEMMA 20.6. *The point f_0 at which φ_i $(i = 1, 2)$ achieves a local minimum (bounded on Δ_δ) is invariant with respect to transformations of the group $\mathfrak{G}_\delta$.*

PROOF. Let $\mathfrak{g} \in \mathfrak{G}_\delta$. Then $\mathfrak{g}$ corresponds to a transformation h of the variables $x_1, \ldots, x_n$ which preserves the simplices $\{S_t\}$, i.e. $h \in H$, where H is the group of affine automorphisms of $\{S_t\}$. From this and from Lemma 20.4 follows the proof of the lemma for local minima. For boundary minima its validity follows from the convexity of Δ_δ and Lemma 20.5.

6. Here we will assume that $\{S_t\}$ is the entire collection of simplices of a star in the L-subdivision of the primitive lattice Γ. Then $\Delta_\delta = \Delta$ is the domain of a single L-type, and if $f \in \Delta$, then $\varphi_1(f) = \theta_n(\Gamma)$ (if $f \notin \Delta$, then $\varphi_1(f) \geqslant \theta_n(\Gamma)$). In this case we obtain from Lemmas 20.3 and 20.6 the following obvious corollaries.

COROLLARY 1 (Barnes and Dickson [20], Theorem 2). *If $f_0 \in \Delta$ is extreme, i.e. there is a local minimum of the density $\theta_n(\Gamma)$ at $f_0 \in \Delta$, then all of the extreme forms of the domain Δ have the form kf_0, where $k > 0$ is arbitrary.*

COROLLARY 2 (Barnes and Dickson [20], Theorem 3). *If the form $f_0 \in \Delta$ is extreme in Δ, then the group of automorphisms of the domain Δ and of the form f_0 coincide.*

We remark that the validity of these statements for boundary minima follows from Lemma 20.6, i.e. those cases where the minimum $\varphi_1(f)$ does not belong to Δ.

7. From formula (19.2) for the function $\theta_n(\Gamma) = \theta_n(f)$ and from the definitions (20.9) of $\varphi_1(f)$ and $\varphi_2(f)$ it follows that every point $f \in \Delta_\delta$ satisfies the inequalities

$$\theta_n(f) \geqslant \varphi_1(f); \qquad \theta_n(f) \geqslant \varphi_2(f). \tag{20.10}$$

Let $\min_L \theta_n(\Gamma)$ be the minimum value of the covering density for the lattices of some L-type one of which belongs to Δ_δ. Then, for such L-types, the inequalities

$$\min_L \theta_n(\Gamma) \geqslant \min \varphi_1(f); \qquad \min_L \theta_n(\Gamma) \geqslant \min \varphi_2(f), \tag{20.11}$$

are fulfilled independently of the positioning of the points in the cone $\boldsymbol{K}$ where $\min \varphi_1(f)$ and $\min \varphi_2(f)$ are attained.

§21. Proof of Lemma 2

1. The radius R of the sphere circumscribed around the simplex S built from the frame $\mathscr{E} = (\bar{e}_1, \ldots, \bar{e}_n)$ may be found with the help of the well-known formula

$$R^2 = -\frac{1}{4} \begin{vmatrix} 0 & \bar{e}_1^2 & \bar{e}_2^2 & \ldots & \bar{e}_n^2 \\ \bar{e}_1^2 & \bar{e}_1^2 & \bar{e}_1\bar{e}_2 & \ldots & \bar{e}_1\bar{e}_n \\ \bar{e}_2^2 & \bar{e}_1\bar{e}_2 & \bar{e}_2^2 & \ldots & \bar{e}_2\bar{e}_n \\ \cdot & \cdot & \cdot & \cdot & \cdot \\ \bar{e}_n^2 & \bar{e}_1\bar{e}_n & \bar{e}_2\bar{e}_n & \ldots & \bar{e}_n^2 \end{vmatrix} (\det f)^{-1}, \tag{21.1}$$

where f is the quadratic form corresponding to $\mathscr{E}$. In our case, for ease in calculating the radii R we modify (21.1) somewhat.

We introduce the following notation for the squares of the lengths of the edges of the simplex S:

$$h_{ij} = (\bar{e}_i - \bar{e}_j)^2; \qquad i, j = 0, 1, \ldots, n; \quad i < j; \quad \bar{e}_0 = 0.$$

Then

$$\bar{e}_i^2 = h_{0i}, \quad \bar{e}_i\bar{e}_j = \frac{1}{2}(-h_{ij} + h_{0i} + h_{0j}), \qquad i, j = 1, \ldots, n, \quad i < j. \tag{21.2}$$

Substituting the values (21.2) for the scalar product $\bar{e}_i\bar{e}_j$ into (21.1), after some transformations we obtain the useful formula

$$R^2 = \frac{(-1)^n}{2^{n+1} \det f} \begin{vmatrix} 0 & h_{01} & h_{02} & \ldots & h_{0n} \\ h_{01} & 0 & h_{12} & \ldots & h_{1n} \\ h_{02} & h_{12} & 0 & \ldots & h_{2n} \\ \cdot & \cdot & \cdot & \cdot & \cdot \\ h_{0n} & h_{1n} & h_{2n} & \ldots & 0 \end{vmatrix}. \tag{21.3}$$

2. We will consider simplices of the repartitioning complex $V_{22}^5(\mathscr{A})$ in the skew state and moreover will assume that this complex corresponds to the component κ_1 in the C-type diagram of Figure 20. We denote by $\{S_t\}$ the set of L-simplices of the star which are homologous to the simplices of Figure 20.

The reduction domain Δ_{sk} for the set of frames of all lattices Γ whose L-subdivision contains the set $\{S_t\}$ is defined by the following inequalities (the first of these is the condition for a skew state; the remaining follow by considering the diagram of the component κ_1 (Figure 15)):

$$\begin{gathered}\rho_{\overline{12}}+\rho_{\overline{13}}+\rho_{\overline{23}}<\rho_{\overline{45}},\quad \rho_{\overline{12}}<\rho_{12},\quad \rho_{\overline{13}}<\rho_{13},\quad \rho_{\overline{23}}<\rho_{23},\\ \rho_{\overline{45}}<\rho_{45},\quad \rho_{\overline{12}}+\rho_{\overline{13}}+\rho_{23},\quad \rho_{\overline{12}}+\rho_{13}+\rho_{\overline{23}},\quad \rho_{12}+\rho_{\overline{13}}+\rho_{\overline{23}}.\end{gathered} \tag{21.4}$$

Obviously the group $\mathfrak{G}_{sk}$ of automorphisms of Δ_{sk} is isomorphic to the group of automorphisms of the diagram of the component κ_1, which is generated by all permutations of vertices 1, 2, 3 of the diagram, the permutations of vertices 4 and 5 and the simultaneous recoloring of vertices 4 and 5.

For the proof of Lemma 2 we will evaluate $\varphi_2(f)$ (see §20) using the set $\{S_t\}$ mentioned in that section, and then will find the minimum of this function. The evaluation of $\varphi_2(f)$ reduces to the determination of the radii R_t of spheres circumscribed around the simplices S_t ($t = 1, \ldots, 6$). As a preliminary to the calculation of the R_t, we make an application of Lemma 20.6.

According to Lemma 20.6, the minimum of $\varphi_2(f)$ is achieved at a point invariant with respect to the group $\mathfrak{G}_{sk}$. We denote the set of such points by Δ_0. Consequently the points $f \in \Delta_0$ satisfy the equalities

$$\begin{gathered}\rho_{12}=\rho_{13}=\rho_{23}=v,\quad \rho_{\overline{12}}=\rho_{\overline{13}}=\rho_{\overline{23}}=\bar{v},\\ \rho_{14}=\rho_{24}=\rho_{15}=\rho_{25}=\rho_{34}=\rho_{35}=\rho_{\overline{14}}=\rho_{\overline{24}}=\rho_{\overline{15}}=\rho_{\overline{25}}=\rho_{\overline{34}}=\rho_{\overline{35}}=u.\end{gathered} \tag{21.5}$$

By considering the one-dimensional skeletons of the simplices $\{S_t\}$, $S'_{1,1}, \ldots, S'_{3,1}$ and $S'_{1,2}, \ldots, S'_{3,2}$ of Figure 20 and by considering the symmetries of the collection of simplices $\{S_t\}$ (Lemma 20.4) of points of Δ_0 where the minimum of $\varphi_2(f)$ occurs we see that the radii of the two spheres circumscribed around the simplices $S'_{1,1}$ and $S'_{1,2}$ coincide, as do the radii of the spheres circumscribed around the simplices $S'_{2,1}$, $S'_{3,1}$, $S'_{2,2}$ and $S'_{3,2}$.

Thus on Δ_0 only two radii of the six will be different. One of them is the radius R_a of the sphere circumscribed around $S'_{1,1}$ and the second is the radius R_b of the sphere circumscribed around $S'_{2,1}$.

From the symmetry of the set $\{S_t\}$ or from (21.5) and (15.7), which amount to the same thing, we get

$$\begin{aligned}&f(l^1)=f(l^2)=f(l^3)=a,\qquad f(l^4)=f(l^5)=b,\\ &f(l^{\overline{12}})=f(l^{\overline{13}})=f(l^{\overline{23}})=c_1,\\ &f(d^1)=f(d^2)=f(d^3)=f(d^{12})=f(d^{13})=f(d^{23})=d_1,\\ &f(d^{14})=f(d^{15})=f(d^{24})=f(d^{25})=f(d^{34})=f(d^{35})=d_2,\\ &f(d^0)=f(d^{45})=h_1,\qquad f(d^4)=f(d^5)=h_2.\end{aligned}$$

Introducing the notation $f(l^{\overline{45}}) = c_2$ and taking advantage of (21.3) and Figure 20, we have

$$R_a^2 = \frac{-1}{2^6 \det f} \begin{vmatrix} 0 & c_1 & c_1 & d_2 & d_1 & d_2 \\ c_1 & 0 & c_1 & d_2 & d_1 & d_2 \\ c_1 & c_1 & 0 & d_2 & d_1 & d_2 \\ d_2 & d_2 & d_2 & 0 & b & c_2 \\ d_1 & d_1 & d_1 & b & 0 & b \\ d_2 & d_2 & d_2 & c_2 & b & 0 \end{vmatrix}, \tag{21.6}$$

$$R_b^2 = \frac{-1}{2^6 \det f} \begin{vmatrix} 0 & c_1 & c_1 & d_2 & d_1 & a \\ c_1 & 0 & c_1 & d_2 & d_1 & a \\ c_1 & c_1 & 0 & d_2 & d_1 & a \\ d_2 & d_2 & d_2 & 0 & b & h_2 \\ d_1 & d_1 & d_1 & b & 0 & h_1 \\ a & a & a & h_2 & h_1 & 0 \end{vmatrix}. \tag{21.7}$$

Taking into consideration the formulas (3.5) relating the coefficients a_{ij} of quadratic forms of the domain R_1 to the coefficients ρ_{ij} and $\rho_{\overline{ij}}$ of (3.4), we get

$$\det f = \frac{1}{4} \begin{vmatrix} q_1 & \rho_{12} - \rho_{\overline{12}} & \rho_{13} - \rho_{\overline{13}} & \cdots & \rho_{1n} - \rho_{\overline{1n}} \\ \rho_{12} - \rho_{\overline{12}} & q_2 & \rho_{23} - \rho_{\overline{23}} & \cdots & \rho_{2n} - \rho_{\overline{2n}} \\ \rho_{13} - \rho_{\overline{13}} & \rho_{23} - \rho_{\overline{23}} & q_3 & \cdots & \rho_{3n} - \rho_{\overline{3n}} \\ \cdots\cdots & \cdots\cdots & \cdots & \cdots & \cdots \\ \rho_{1n} - \rho_{\overline{1n}} & \rho_{2n} - \rho_{\overline{2n}} & \rho_{3n} - \rho_{\overline{3n}} & \cdots & q_n \end{vmatrix}, \tag{21.8}$$

where

$$q_i = \sum_{\substack{j=1 \\ (j \neq i)}}^{n} (\rho_{ij} + \rho_{\overline{ij}}), \qquad i = 1, \ldots, n.$$

On the basis of (21.5) we intoduce the new notation

$$\rho = \rho_{45} + 3u, \qquad \delta = \rho_{\overline{45}} + 3u, \qquad \alpha = v + u, \qquad \beta = \bar{v} + u. \tag{21.9}$$

According to (21.4) the set Δ_0 is determined by the inequalities $\rho > \delta$, $\alpha > \beta$ and $\delta > 3\beta$, and $\det f$ (21.8) may be evaluated at points of Δ_0 using the formula

$$\det f = 4\rho\delta\alpha(\alpha + 3\beta)^2. \tag{21.10}$$

Using (15.7) and the notation of (21.9), we calculate the squares of the lengths of the contiguity vectors a, b, c_1, c_2, d_1, d_2, h_1 and h_2. We get

$$a = 2(\alpha + \beta), \qquad b = \rho + \delta, \qquad c_1 = 2(\alpha + 3\beta), \qquad c_2 = 4\delta,$$
$$d_1 = \rho + \alpha + 2\beta, \qquad d_2 = \delta + \alpha + 2\beta, \qquad h_1 = \rho + 3\alpha, \qquad h_2 = \delta + 3\alpha.$$

Substituting these quantities and the value (21.10) of $\det f$ into (21.6) and (21.7), we obtain, after some computation,

$$
\begin{aligned}
R_a^2 &= \frac{1}{4\rho\delta}\left[\alpha(\rho^2+\delta^2)+3\rho\delta^2+3\rho(\alpha+2\beta)^2-4\rho\delta(\alpha+3\beta)\right],\\
R_b^2 &= \frac{3}{8\rho\delta\alpha}\left[(\alpha-\beta)^2(\rho^2+\delta^2)+2(\rho+\delta)\alpha\left(3\beta^2+\frac{\rho\delta}{3}\right)\right].
\end{aligned}
\tag{21.11}
$$

3. Since we are only interested in $\min \varphi_2(f)$, it is sufficient to assign values to the coefficients p_t of (20.5) only for the cases where $f \in \Delta_0$. The function $\varphi_2(f)$ on Δ_0 has the expression

$$\varphi_2(f) = \Omega_5 \sqrt{\frac{(R_a^2+R_b^2)^5}{2^5 \det f}}.$$

Here the values of R_a^2 and R_b^2 are obtained from (21.11), and the value of $\det f$, from (21.10). The task of calculating $\min \varphi_2(f)$ has been reduced to the calculation of the minimum of a function which essentially depends on only three variables:

$$\varphi(\rho:\delta:\alpha:\beta) = \frac{(R_a^2+R_b^2)^5}{2^5 \det f}.$$

Here we employed a computer and obtained the following result: the unique minimum of the function $\varphi(\rho:\delta:\alpha:\beta)$ is achieved at the point

$$\rho = 1.60573\ldots,\quad \delta = 1.24347\ldots,\quad \alpha = 1.00000\ldots,\quad \beta = 0.288387\ldots$$

(uniqueness of the minimum follows from Lemma 20.4) and is equal to 0.167090403 (accuracy to within 10^{-9}).

According to the second of formulas (20.12), the L-types whose L-subdivisions contain the simplices $\{S_t\}$ considered in this section, i.e. the nonkey L-types of R_1, have a minimum value of the density $\theta_5(\Gamma)$ which is not less than

$$\min \varphi_2(f) = \Omega_5 \sqrt{0.1670904} = \Omega_5\, 0.408\ldots.$$

Thus

$$\min \varphi_2(f) > \theta_5^0 = \Omega_5\, 0.403\ldots,$$

and this completes the proof of Lemma 2.

§22. Investigation of the covering densities of the key L-types 1–7 of the domain R_1 (Proof of Lemma 4)

1. The L-type referred to in our work for $n = 5$ as the key L-type 1 was investigated for all $n \geqslant 5$ by Barnes and Trenerry [23], [39]. Moreover, they obtained the value of the covering density for lattices of this type [23]. Below we describe our own calculation for the covering density of L-type 1, but draw the reader's attention to their work [23].

According to the results of §20.6 (Corollaries 1 and 2), the minimum density will occur on the set τ_1 of points of Δ_1 corresponding to the diagram $\mathfrak{S}_1$ of L-type 1 which are fixed by the group of automorphisms of this domain. Since the group of automorphisms of Δ_1 is isomorphic to the group of all permutations of the numbers $1, \ldots, 5$, the elements of τ_1 have the coordinates

$$\rho_{12} = \ldots = \rho_{45} = \alpha, \quad \rho_{\overline{12}} = \ldots = \rho_{\overline{45}} = \beta, \quad \alpha < \beta. \tag{22.1}$$

From (22.1) and from the description of the L-simplices of lattices of this type (§15, Example 1), it follows that all of the L-simplices are congruent for lattices belonging to τ_1. Consequently the radii of their circumscribed spheres are equal. Taking advantage of (15.7) and (21.2), we find that the square of the radius of these spheres, which is equal to the square of the covering radius, is given by

$$R^2 = \frac{19\alpha^2 + 20\beta^2 + 33\alpha\beta}{2(3\alpha + 5\beta)}.$$

By (19.1) the covering density is

$$\theta(\alpha:\beta) = \Omega_5 \frac{(19\alpha^2 + 20\beta^2 + 33\alpha\beta)^{5/2}}{8\sqrt{\alpha}\,(3\alpha + 5\beta)^{9/2}}. \tag{22.2}$$

The minimum of (22.2) with respect to the variable $x = \alpha:\beta$ is easy to calculate and is equal to $\Omega_5 \cdot 0.4237\ldots$. The minimum is achieved at the point $\alpha:\beta = 1:2.2209\ldots$, which belongs to τ_1. This is the solution of the problem of the minimum covering density for the L-type 1.

2. *The key L-type* 5. The domain Δ_5 is defined by the inequalities of the diagram $\mathfrak{S}_5$ and by the collection of inequalities

$$\rho_{\overline{ij}} < \rho_{\overline{km}} + \rho_{\overline{kp}} + \rho_{\overline{mp}} \quad (i, j, k, m, p = 1, \ldots, 5)$$

defining the normal state of all of the repartitioning complexes of C-type 5. Its group of automorphisms is equal to the group of automorphisms of Δ_1. The set $\tau_5 \subset \Delta_5$, which is fixed with respect to this group, satisfies all of the equalities (22.1) and the inequality $\alpha > \beta$.

By considering the description of the L-simplices of the type under consideration given in Example 2 of §15 we discover that they separate into three sets of mutually congruent simplices. Each simplex of the first set (the simplex $S_{1\overline{2},3\overline{4}}$ is a representative) appears in two of the repartitioning complexes $V^5_{22}(\mathscr{A})$; each simplex of the second set ($S_{1\overline{2},\overline{3}4}$ is a representative) appears in only one such complex, and each simplex of the third set ($S_{\overline{1}2,3\overline{4}}$ is a representative) does not appear in any of the repartitioning complexes.

Calculating the radii of the circumscribed spheres for each of these three sets of simplices, we get

$$R_1^2 = \frac{15x^2 + 25x + 32}{2(5 + 3x)}\beta, \quad R_2^2 = \frac{12x^2 + 32x + 27 + \frac{1}{x}}{2(5 + 3x)}\beta, \quad R_3^2 = \frac{11x^2 + 33x + 28}{2(5 + 3x)}\beta,$$

where $x = \alpha:\beta$.

Since $x > 1$ on τ_5, we have $\max_{t=1,2,3} R_t = R_1$, and the covering density is given by

$$\theta(x) = \Omega_5 \frac{R_1^5}{\sqrt{\det f}} = \frac{\Omega_5}{8} \frac{(15x^2 + 25x + 32)^{5/2}}{x^{1/2}(5 + 3x)^{9/2}}.$$

After a calculation we find that $\min \theta(x) = 0.437\ldots$ and that this minimum is achieved at $\alpha : \beta = 1.8832\ldots$. Obviously $\min \theta(x)$ is the minimum covering density for the L-type 5.

3. *The key L-types* 2 *and* 6. As is obvious from diagrams $\mathfrak{S}_2$ and $\mathfrak{S}_6$, all of the repartitioning complexes of these C-types are in a normal state. Therefore Δ_2 and Δ_6 and the diagrams of the key L-types corresponding to them have identical automorphism groups. This group is isomorphic to the group generated by all permutations of the vertices 3, 4, 5 and permutations of vertices 1 and 2 of diagrams $\mathfrak{S}_2$ and $\mathfrak{S}_6$.

From this it follows that the sets $\tau_2 \subset \Delta_2$ and $\tau_6 \subset \Delta_6$ which are invariant with respect to these groups satisfy the equalities

$$\begin{aligned} &\rho_{13} = \rho_{14} = \rho_{15} = \rho_{23} = \rho_{24} = \rho_{25} = u, \qquad \rho_{34} = \rho_{35} = \rho_{45} = v, \\ &\rho_{\overline{13}} = \rho_{\overline{14}} = \rho_{\overline{15}} = \rho_{\overline{23}} = \rho_{\overline{24}} = \rho_{\overline{25}} = \bar{u}, \qquad \rho_{\overline{34}} = \rho_{\overline{35}} = \rho_{\overline{45}} = \bar{v}. \end{aligned} \tag{22.3}$$

In place of the six parameters $u, v, \bar{u}, \bar{v}, \rho_{12}$ and $\rho_{\overline{12}}$ which determine in a nonunique way (as with the parameters ρ_{ij} and $\rho_{\overline{ij}}$) the lattices of τ_2 and τ_6, we introduce the new parameters

$$\rho = \rho_{12} + \frac{3}{2}(u + \bar{u}), \qquad \bar{\rho} = \rho_{\overline{12}} + \frac{3}{2}(u + \bar{u}), \qquad b = v + \frac{u + \bar{u}}{2},$$
$$l = 2(u + v + \bar{u} + \bar{v}), \qquad a = u - \bar{u}.$$

which determine these lattices in a one-to-one fashion.

Using (15.7) and (22.3), we find the squares of the lengths of the contiguity vectors of Δ_2 and Δ_6 for the sets τ_2 and τ_6. We will denote the contiguity vectors and the squares of their lengths by the same sign throughout the remainder of §22. We have

$$\begin{aligned} &l^3 = l^4 = l^5 = l, \quad l^1 = l^2 = \rho + \bar{\rho}, \quad l^{\overline{12}} = 4\bar{\rho}, \quad l^{12} = 4\rho, \\ &l^{13} = l^{14} = l^{15} = l^{23} = l^{24} = l^{25} = \rho + \bar{\rho} + l + 2a, \\ &l^{\overline{13}} = l^{\overline{14}} = l^{\overline{15}} = l^{\overline{23}} = l^{\overline{24}} = l^{\overline{25}} = \rho + \bar{\rho} + l - 2a, \\ &l^{34} = l^{35} = l^{45} = l + 4b, \quad l^{\overline{34}} = l^{\overline{35}} = l^{\overline{45}} = 3l - 4b, \\ &d^3 = d^4 = d^5 = \rho + l - b + a, \quad d^{34} = d^{35} = d^{45} = \rho + l - b - a, \\ &d^{13} = d^{14} = d^{15} = d^{23} = d^{24} = d^{25} = \bar{\rho} + l - b, \\ &d^1 = d^2 = \bar{\rho} + 3b, \quad d^0 = \rho + 3b + 3a, \quad d^{12} = \rho + 3b - 3a. \end{aligned}$$

We remark that elements of τ_2 and τ_6 satisfy

$$\begin{aligned} &\tau_2\colon \bar{\rho} < \rho, \quad a < 0, \quad 3b > \bar{\rho}, \quad 4b - l < 0, \quad 2b + a > 0, \quad \bar{\rho} > b - a, \\ &\tau_6\colon \rho < \bar{\rho}, \quad a > 0, \quad 4b - l > 0, \quad l - 2b - a > 0, \quad \rho - 3b - {}^3/_2 l > 0. \end{aligned}$$

For quadratic forms from either τ_2 or τ_6 we have the general formula

$$\det f = \bar{\rho}(3l-4b)^2\left(\rho b - \frac{3}{4}a^2\right).$$

We choose from among the mutually nonhomologous L-simplices of lattices of type 2 the simplex

$$S_2 = (0,\ l^5,\ l^{15},\ d^3,\ d^{34},\ -d^{15})$$

of block (15, 24), and from those of type 6 the simplex

$$S_6 = (0,\ -l^1,\ -l^{\overline{15}},\ d^{14},\ d^1,\ d^{12})$$

of block $(\overline{15}, \overline{24})$. We denote by R_ω ($\omega = 2, 6$) the radius of the sphere circumscribed around S_ω. Obviously, for the domains τ_ω ($\omega = 2, 6$), R_ω is the radius of a sphere circumscribed around each of several (at least six) mutually nonhomologous L-simplices.

Making use of (21.3), we find that

$$R_2^2 = \frac{1}{4}\left[\rho + \bar{\rho} + l - b + 4a + \frac{9}{4}\frac{a^2}{\bar{\rho}} + \frac{(l-2a)^2}{3l-4b}\right],$$

$$R_6^2 = \frac{1}{4}\left[\rho + \bar{\rho} + l - b - 4a + \frac{9}{4}\frac{a^2}{\bar{\rho}} + \frac{(l+2a)^2}{3l-4b}\right],$$

and we need to find the minima of the functions

$$\Phi_\omega = \Phi_\omega(\rho : \bar{\rho} : l : b : a) = \Omega_5 \frac{R_\omega^5}{\sqrt{\det f}} \qquad (\omega = 2,6).$$

Since Φ_2 transforms into Φ_6 under the change of variables $a \longrightarrow -a$, we need only search for the minimum of Φ_2.

Our functions Φ_ω ($\omega = 2, 6$) belong to that class of functions which were considered in §20 and denoted there by $\varphi_1(f)$. It follows from Lemma 20.3 that the functions Φ_ω have one and only one minimum on the cone K. As was shown in §20, these minima are lower bounds for the covering densities for types 2 and 6 respectively.

We will find the critical points of Φ_2. In the usual way we find that these points must satisfy the relations

$$\frac{1}{\bar{\rho}} = \frac{32b^2 - 4ab - 9al - 16bl}{9ab(3l-4b)}, \qquad \rho = \frac{3}{4}\frac{a^2}{b} + \frac{(3l-4b)^2 + (l-2a)(3l-8b+6a)}{6(3l-4b)},$$

$$\frac{18ab(3l-4b)}{32b^2 - 4ab - 9al - 16bl} = 2l - 5a - \frac{7}{3}b - \frac{3}{4}\frac{a^2}{b} + \frac{7(l-2a)(3a-2b)}{3(3l-4b)},$$

$$2l^2 - 9bl + 4b^2 + 8ab - 5a^2 + \frac{9}{4}\frac{a^2 l}{b} = 0.$$

Investigation of these equations, for example when $b = 1$, yields the following values of the variables at the critical point ξ:

$$\rho = 1.878\ldots, \qquad \bar{\rho} = 1.956\ldots, \qquad l = 4.339\ldots, \qquad a = -0.4427\ldots.$$

From the above discussion it follows that ξ is a minimum point of Φ_2 and that the value at the minimum, namely $\Phi_2(\xi) = \Omega_5 \cdot 0.415\ldots$, is a lower bound to the covering density. Thus for arbitrary lattices Γ of the key L-types 2 and 6 we have $\theta_5(\Gamma) > \Omega_5 \cdot 0.415$.

4. *The key L-types* 3 *and* 7. By recoloring vertices 1 and 2 of diagrams $\mathfrak{S}_3$ and $\mathfrak{S}_7$ we obtain respectively the diagrams $\mathfrak{S}'_3$ and $\mathfrak{S}'_7$. The automorphism group of both of these is generated by the permutations of each of the pairs of vertices 1, 2 and 3, 5 and the exchange of these pairs. This group is isomorphic to both the group of automorphisms of the domain Δ_3 of the key L-type 3 corresponding to $\mathfrak{S}'_3$ and the group of automorphisms of the domain Δ_7 of the key L-type 7 corresponding to $\mathfrak{S}'_7$.

Elements belonging to the sets $\tau_3 \subset \Delta_3$ and $\tau_7 \subset \Delta_7$ are fixed under the automorphism groups of their respective domains Δ_3 and Δ_7 and satisfy the following equalities:

$$\begin{aligned} &\rho_{13}=\rho_{15}=\rho_{23}=\rho_{24}=u, && \rho_{14}=\rho_{24}=\rho_{34}=\rho_{45}=v, && \rho_{12}=\rho_{35}=w, \\ &\rho_{\overline{13}}=\rho_{\overline{15}}=\rho_{\overline{23}}=\rho_{\overline{25}}=u, && \rho_{\overline{14}}=\rho_{\overline{24}}=\rho_{\overline{34}}=\rho_{\overline{45}}=\bar v, && \rho_{\overline{12}}=\rho_{\overline{35}}=\bar w. \end{aligned} \tag{22.4}$$

As in subsection 3, we transform to new parameters which are in a one-to-one relation with the coefficients of the quadratic form:

$$\rho=w+u+\bar u+\frac{v+\bar v}{2},\qquad \bar\rho=\bar w+u+\bar u+\frac{v+\bar v}{2},\qquad m=v+\bar v,$$

$$\varepsilon=u-\bar u,\qquad \delta=v-\bar v.$$

In terms of these parameters the domain τ_3 is given by the inequalities $\bar\rho>\rho$, $\rho-\epsilon>m-\delta/2$, $\epsilon>0$, $\delta>0$ and $m-\delta>0$, and τ_7 by the inequalities $\rho>\bar\rho$, $\bar\rho+\epsilon>m+\delta/2$, $\epsilon<0$, $\delta<0$ and $m+\delta>0$.

Making use of (15.7), we find the squares of the lengths of the contiguity vectors of τ_3 and τ_7:

$$\begin{aligned} &l^1=l^2=l^3=l^5=\rho+\bar\rho,\qquad l^4=4m,\qquad l^{12}=l^{35}=4\rho,\\ &l^{\overline{13}}=l^{\overline{15}}=l^{\overline{23}}=l^{\overline{25}}=2(\rho+\bar\rho-\varepsilon),\\ &l^{\overline{14}}=l^{\overline{24}}=l^{\overline{34}}=l^{\overline{45}}=\rho+\bar\rho+4m-2\delta,\\ &l^{\overline{12}}=l^{\overline{35}}=4\bar\rho,\qquad l^{13}=l^{15}=l^{23}=l^{25}=2(\rho+\bar\rho+\varepsilon),\\ &l^{14}=l^{24}=l^{34}=l^{45}=\rho+\bar\rho+4m+2\delta, \end{aligned}$$

$$\begin{aligned} &d^0=2(\rho+\varepsilon)+m+2\delta, && d^4=2(\rho+\varepsilon)+m-2\delta,\\ &d^1=d^2=d^3=d^5=\rho+\bar\rho+m+\delta, && d^{12}=d^{35}=2(\rho-\varepsilon)+m,\\ &d^{13}=d^{15}=d^{23}=d^{25}=2\bar\rho+m, && d^{14}=d^{24}=d^{34}=d^{45}=\rho+\bar\rho+m-\delta. \end{aligned}$$

The determinants of quadratic forms belonging to either τ_3 or τ_7 are found by the formula

$$\det f=8\bar\rho^2(\rho-\varepsilon)[2m(\rho+\varepsilon)-\delta^2].$$

From the set of mutually nonhomologous L-simplices corresponding to points of τ_3 we consider simplices from blocks (12, 35) and $(\overline{13}, \overline{25})$. The simplices of block (12, 35) are mutually congruent, the squares of the radii of their circumscribed spheres being

$$R_1^2=\frac{1}{4}\left[2\bar\rho+2\frac{\rho^2}{\bar\rho}+m+\frac{\delta^2}{2(\rho-\varepsilon)}\right].$$

From block $(\overline{13}, \overline{25})$ we select the simplices

$$(0,\ l^1,\ l^{\overline{13}},\ d^{23},\ -d^{14},\ d^{35}) \qquad \text{and} \qquad (0,\ -l^3,\ l^{\overline{13}},\ d^{23},\ d^3,\ d^{35})$$

and find the radii of their circumscribed spheres:

$$R_2^2 = \frac{1}{2}\left[\rho + \bar\rho + \frac{m}{2} - \varepsilon + \frac{1}{\bar\rho}\left(\varepsilon - \frac{\delta}{2}\right)^2\right],$$

$$R_3^2 = \frac{1}{2}\left[\rho + \bar\rho + \frac{m}{2} - \varepsilon + \frac{1}{\bar\rho}\left(\varepsilon + \frac{\delta}{2}\right)^2\right].$$

We will search for the minimum of the function

$$\Phi_3 = \Phi_3(\rho : \bar\rho : m : \varepsilon : \delta) = \Omega_5 \frac{\left(\max\limits_{i=1,2,3} R_i\right)^5}{\sqrt{\det f}},$$

which belongs to the class of functions $\varphi_1(f)$ considered in §20. We find the following critical "point" (ray) $\rho : \bar\rho : m : \epsilon : \delta$ of the function Φ_3 defined on K:

$$\rho = 1,\quad \bar\rho = \sqrt{13} - 2,\quad m = {}^1/_9\ (5\sqrt{13} - 8),\quad \varepsilon = \sqrt{13} - 3,\ \delta = 0.$$

At this point $\Phi_3 = \Omega_5 \cdot 0.4243\ldots$. According to Lemma 20.3 this point is the unique minimum of Φ_3, and the value of Φ_3 at this point gives a lower bound to the covering density for lattices of type 3.

Type 7 is more difficult. Here the formulas we obtained were too awkward for "hand" calculation and the results were obtained by computer. For the domain τ_7 there are 12 mutually inequivalent simplices. They are

$$\begin{array}{ll}
S_{13,45} = (0,\ l^1,\ l^{13},\ d^2,\ d^{24},\ -d^{13}), & S_{31,45} = (0,\ l^3,\ l^{13},\ d^2,\ d^{24},\ -d^{13}),\\
S_{13,54} = (0,\ l^1,\ l^{13},\ d^2,\ d^{25},\ -d^{13}), & S_{31,54} = (0,\ l^3,\ l^{13},\ d^2,\ d^{25},\ -d^{13}),\\
S_{1\bar2,34} = (0,\ l^1,\ l^{\overline{12}},\ d^{25},\ -d^{14},\ -d^1), & S_{\bar12,34} = (0,\ -l^2,\ l^{\overline{12}},\ d^{25},\ -d^{14},\ -d^1),\\
S_{1\bar2,43} = (0,\ l^1,\ l^{\overline{12}},\ d^{25},\ -d^{13},\ -d^1), & S_{\bar12,43} = (0,\ -l^2,\ l^{\overline{12}},\ d^{25},\ -d^{13},\ -d^1),\\
S_{\bar12,\bar35} = (0,\ l^1,\ l^{\overline{12}},\ d^{23},\ -d^{14},\ d^{25}), & S_{1\bar2,\bar35} = (0,\ -l^2,\ l^{\overline{12}},\ d^{23},\ -d^{14},\ d^{25}),\\
S_{31,25} = (0,\ l^3,\ l^{13},\ d^4,\ d^{24},\ -d^{13}), & S_{13,25} = (0,\ l^1,\ l^{13},\ d^4,\ d^{24},\ -d^{13}).
\end{array}$$

For a bound on the minimum of the function

$$\Phi_7(\bar\rho : \rho : m : \varepsilon : \delta) = \Omega_5 \left(\max_{i=1,2,\ldots,12} R_i\right)^5 (\det f)^{-1/2}$$

we found the minimum of the function

$$\Omega_5 \left(\frac{1}{12}\sum_{i=1}^{12} R_1^2\right)^{5/2} (\det f)^{-1/2}.$$

It occurred at the point

$$1 : 1.385\ldots : 0.5979\ldots : -0.2540\ldots : -0.2408\ldots$$

and was equal to $\Omega_5 \cdot 0.4136\ldots$. Thus for this type we have $\Phi_7 \geqslant \Omega_5 \cdot 0.4136\ldots$.

5. *The key L-type* 4. The automorphism group of key type 4, corresponding to the diagram $\mathfrak{S}_4$, as is clear from this diagram, is isomorphic to the group generated by permutations of the form

$$\begin{pmatrix}1&2&3&4&5\\2&3&4&5&1\end{pmatrix}\quad\text{and}\quad\begin{pmatrix}1&2&3&4&5\\1&5&4&3&2\end{pmatrix}.$$

Elements belonging to $\tau_4 \subset \Delta_4$ are fixed under the automorphism group of Δ_4 and satisfy the following equalities:

$$\rho_{12}=\rho_{23}=\rho_{34}=\rho_{45}=\rho_{15}=u,\qquad \rho_{\bar{1}\bar{2}}=\rho_{\bar{2}\bar{3}}=\rho_{\bar{3}\bar{4}}=\rho_{\bar{4}\bar{5}}=\rho_{\bar{1}\bar{5}}=\bar{u},$$
$$\rho_{13}=\rho_{14}=\rho_{24}=\rho_{25}=\rho_{35}=v,\qquad \rho_{\bar{1}\bar{3}}=\rho_{\bar{1}\bar{4}}=\rho_{\bar{2}\bar{4}}=\rho_{\bar{2}\bar{5}}=\rho_{\bar{3}\bar{5}}=\bar{v}.$$

We change to the new parameters $l = 2(u + v + \bar{u} + \bar{v})$, $\kappa = 2(u - \bar{u})$ and $\lambda = 2(v - \bar{v})$, which are in one-to-one correspondence with the coefficients of the quadratic forms. Using (15.7), we find the squares of the lengths of the contiguity vectors for elements belonging to τ_4:

$$l^1=l^2=\ldots=l^5=l,\qquad l^{\bar{1}\bar{2}}=l^{\bar{2}\bar{3}}=l^{\bar{3}\bar{4}}=l^{\bar{4}\bar{5}}=l^{\bar{1}\bar{5}}=2l-\varkappa,$$
$$l^{13}=l^{14}=l^{24}=l^{25}=l^{35}=2l+\lambda,$$
$$d^0=\frac{5}{4}(l+\varkappa+\lambda),\qquad d^1=d^2=\ldots=d^5=\frac{5}{4}l+\frac{\varkappa+\lambda}{4},$$
$$d^{12}=d^{23}=d^{34}=d^{45}=d^{15}=\frac{5}{4}l+\frac{\varkappa-3\lambda}{4},$$
$$d^{13}=d^{14}=d^{24}=d^{25}=d^{35}=\frac{5}{4}l+\frac{\lambda-3\varkappa}{4}.$$

On τ_4, where $l > 0$, $\kappa > 0$ and $\lambda < 0$, we have

$$\det f=\frac{1}{4}(l+\varkappa+\lambda)\left[\left(l-\frac{\varkappa}{2}\right)\left(l-\frac{\lambda}{2}\right)-\frac{(\varkappa-\lambda)^2}{4}\right]^2.$$

We consider the simplices in the blocks of the L-subdivision of lattices of τ_4. Due to the symmetries occurring we conclude that there are in total only 8 distinct radii for the spheres circumscribed around each of these L-simplices. These are the radii $R_1, \ldots, R_8$ of the spheres circumscribed, for example, around the simplices

$$S_{1\bar{2},\bar{3}4},\ S_{\bar{1}2,\bar{3}4},\ S_{1\bar{2},3\bar{4}},\ S_{13,\bar{4}\bar{5}},\ S_{13,\bar{4}5},\ S_{13,24},\ S_{31,24},\ S_{13,42}.$$

In order to estimate the minimum of the function

$$\Phi_4(l:\varkappa:\lambda)=\Omega_5(\max_{i=1,2,\ldots,8}R_i)^5(\det f)^{-1/2},$$

we consider the function

$$\begin{aligned}\Psi_4(l:\varkappa:\lambda)&=\Omega_5R_5^5(\det f)^{-1/2}\\&=\Omega_5 2^{-15}(\det f)^{-3}\Big[9l^6+2l^5(\lambda-\varkappa)-\frac{31}{4}l^4(\varkappa^2+\lambda^2)\\&+l^3\left(\varkappa^3+5\varkappa^2\lambda+8\varkappa\lambda^2-\frac{1}{2}\lambda^3\right)+\frac{l^2}{16}(29\varkappa^4-122\varkappa^3\lambda\\&+39\varkappa^2\lambda^2-98\varkappa\lambda^3+5\lambda^4)+\frac{l}{8}(5\varkappa^5+39\varkappa^4\lambda+15\varkappa^3\lambda^2\\&+77\varkappa^2\lambda^3-9\varkappa\lambda^4+\lambda^5)+\frac{1}{16}(\varkappa^6-8\varkappa^5\lambda+28\varkappa^4\lambda^2\\&-38\varkappa^3\lambda^3+4\lambda^4\varkappa^2+\lambda^6)\Big]^{5/2}.\end{aligned}$$

The minimum of $\Psi_4(l:\kappa:\lambda)$ is achieved at the point

$$1:0{,}1778\ldots: -0{,}1902\ldots$$

and is equal to $\Omega_5 \cdot 0.4279 \ldots$. Therefore, for this type we have

$$\Phi_4 \geqslant \Omega_5 \cdot 0{,}4279.$$

§23. On a single bound for the covering density for each of the L-type n-dimensional lattices

1. Let $\bar{l}_1, \ldots, \bar{l}_\sigma$ ($\sigma = 2^n - 1$) be contiguity vectors for a (general) L-type, the coordinates being given with respect to a fundamental frame to which the quadratic form $f(x_1, \ldots, x_n)$ corresponds. The square of the lengths of each of these vectors $\bar{l}_p = (l_1^{(p)}, \ldots, l_n^{(p)})$ ($p = 1, \ldots, \sigma$) can be found by the formula

$$f(\bar{l}_p) = f(l_1^{(p)}, \ldots, l_n^{(p)}) = \sum_{i,j=1}^{n} a_{ij} l_i^{(p)} l_j^{(p)}. \tag{23.1}$$

We denote by $S_1, \ldots, S_q$ the entire set of mutually nonhomologous L-simplices of a lattice of this type. We denote the vertices of S_t ($t = 1, \ldots, q$) by $A_0^{(t)}, \ldots, A_n^{(t)}$. Since each of the edges $A_k^{(t)}A_m^{(t)}$ of the simplices $S_1, \ldots, S_q$ corresponds to one of the contiguity vectors of this type, the squares of the lengths of the edges $A_k^{(t)}A_m^{(t)}$ may be calculated by (23.1).

We consider the quantity

$$M(\Gamma) = \frac{1}{q(n+1)} \sum_{t=1}^{q} \sum_{\substack{k,m=0 \\ (k<m)}}^{n} (A_k^{(t)}A_m^{(t)})^2 = \frac{1}{q(n+1)} \sum_{i,j=1}^{n} c_{ij} a_{ij}, \tag{23.2}$$

where the coefficients $c_{ij} = c_{ji}$, as was mentioned above, are positive integers characterizing the given L-type. We have

THEOREM 23.1. *The minimum value of the covering density $\theta_n(\Gamma)$ for lattices of a given general L-type is not less than the quantity*

$$M_n = \Omega_n \left[\frac{n}{q(n+1)^2}\right]^{n/2} \sqrt{\det(c_{ij})}, \tag{23.3}$$

where q is the number of mutually nonhomologous L-simplices of lattices of this type and $\det(c_{ij})$ *is the determinant of the coefficients appearing in* (23.2).

The proof of this theorem will be given in subsection 4, summarizing the lemmas appearing in subsections 2 and 3. We will finish this subsection with some comments of an historical character.

One of the ideas on which the estimate (23.3) for the covering density is based was drawn from the famous theorem of Steiner on the central moment of inertia and in the context of lattice coverings was first applied by S. S. Ryškov in his unpublished solution of the problem of the least dense lattice covering for four-dimensional lattices of Voronoĭ's first type [35]. Later the "method of central moments" was used in the work of Gameckiĭ [52], Delone and Ryškov [19] and Ryškov [22]. Moreover, for $n = 4$ the lower bound (23.3) of the covering density found with the help of the "method of central moments" yields a higher lower bound for the three L-types of the four-dimensional lattices (see Theorem T_n) than does the method used to obtain the proof of Theorem T_n for $n = 4$.

Our estimate (23.3) allows a quick comparison among a given set of L-types; for $n \leqslant 4$ such an investigation quickly yields a proof of Theorem T_n (Examples 1 and 2). For the proof of Theorem T_5 the estimate (23.3) does not play an important role. It reduces the number of L-types requiring consideration from 221 to 74.

2. In E^n let S be an n-dimensional simplex with vertices $A_0, \ldots, A_n$, and let B be some point. The moment $M(B)$ of the simplex, or, more precisely, the moment of the vertices of the simplex with respect to the point B, is given by

$$M(B) = \sum_{i=0}^{n} BA_i^2.$$

If B is the center of mass of S, the moment is called central. We will denote this moment by M_S.

Proposition 23.2. *Let $S = A_0A_1 \cdots A_n$ be an n-dimensional simplex, O its center of mass, M_S its central moment, R its radius and C the center of the sphere circumscribed around S. Then*

$$M_S = \frac{1}{n+1} \sum_{\substack{i,j=0 \\ (i<j)}}^{n} A_iA_j^2, \tag{23.4}$$

$$R^2 = \frac{M_S}{n+1} + CO^2 = \frac{1}{(n+1)^2} \sum_{\substack{i,j=0 \\ (i<j)}}^{n} A_iA_j^2 + CO^2. \tag{23.5}$$

Proof. Let $A_1, \ldots, A_m$ and B be $m+1$ arbitrary points in E^n, and let O be the center of mass of the system $A_1, \ldots, A_m$. Then

$$\sum_{p=1}^{m} BA_p^2 = \sum_{p=1}^{m} OA_p^2 + mBO^2. \tag{23.6}$$

In order to prove this equality we square the vector equality $\overline{BA_p} = \overline{OA_p} + \overline{BO}$, $p = 1, \ldots, m$, and add. We obtain

$$\sum_{p=1}^{m} BA_p^2 = \sum_{p=1}^{m} OA_p^2 + 2\overline{BO}\left(\sum_{p=1}^{m} \overline{OA_p}\right) + mBO^2.$$

According to the definition of the center of mass

$$\sum_{p=1}^{m} \overline{OA}_p = \bar{0},$$

and we have (23.6).

If $M(B)$ is the moment of the simplex S with respect to B, then by (23.6) and the definiton of the central moment M_S we have

$$M(B) = M_S + (n+1)\, BO^2. \tag{23.7}$$

Assuming that the points $A_0, \ldots, A_n$ are vertices of S and taking advantage of (23.7), we get

$$M(A_0) = \sum_{i=0}^{n} A_0A_i^2 = M_S + (n+1)\, OA_0^2,$$

$$M(A_1) = \sum_{i=0}^{n} A_1A_i^2 = M_S + (n+1)\, OA_1^2,$$

$$\cdots\cdots\cdots\cdots\cdots\cdots$$

$$M(A_n) = \sum_{i=0}^{n} A_nA_i^2 = M_S + (n+1)\, OA_n^2.$$

After adding these equalities we have

$$2 \sum_{\substack{i,\, j=0 \\ (i<j)}}^{n} A_iA_j^2 = (n+1)\, M_S + (n+1) \sum_{i=1}^{n} OA_i^2 = 2(n+1)\, M_S,$$

from which (23.4) follows.

We make use of (23.7) again for the calculation of $M(C)$:

$$M(C) = \sum_{i=0}^{n} CA_i^2 = M_S + (n+1)\, CO^2.$$

Taking into account that $R = |\overline{CA_i}|$, we have

$$(n+1) R^2 = M_S + (n+1)\, CO^2,$$

and from (23.4) both of the equalities (23.5) quickly follow. This finishes the proof of the proposition.

3. Assume that we have found the central moments for all of the mutually non-homologous L-simplices $S_1, \ldots, S_q$ for some L-type using formula (23.4).

The arithmetical mean $q^{-1}\Sigma_{t=1}^{q} M_{S_t}$ of these moments $M_{S_1}, \ldots, M_{S_q}$ is equal to the quantity $M(\Gamma)$ of (23.2) and is the "central moment of inertia of the lattice Γ". Considering $M(\Gamma)$ to be a function of the variables a_{ij}, we will show that at its minimum on K the formula $\det f = \det(a_{ij})$ holds.

PROPOSITION 23.3. *The minimum of the function*

$$M(\Gamma) = \frac{1}{q(n+1)} \sum_{i,\, j=1}^{n} c_{ij} a_{ij}$$

on the equidiscriminant surface $\det(a_{ij}) = d_0$ *is equal to*

$$\frac{n}{q(n+1)} \sqrt[n]{d_0 \cdot \det(c_{ij})}$$

and it occurs at a unique point on this surface. The coordinates a_{ij}^* of this point are the elements of the inverse of the matrix (c_{ij}) multiplied by $[d_0 \det(c_{ij})]^{-1/n}$.

PROOF. In order to find the minimum we will use the method of Lagrange. We take the function

$$F(a_{ij}) = \sum_{i, j=1}^{n} c_{ij} a_{ij} - \lambda [\det(a_{ij}) - d_0]$$

(for ease in computation we have dropped the quantity $1/q(n+1)$ from the right-hand side of (23.2)) and find its partial derivatives with respect to the variables a_{ij}. We have

$$\begin{aligned} \frac{\partial F}{\partial a_{ii}} &= c_{ii} - \lambda A_{ii} \qquad (i = 1, \ldots, n), \\ \frac{\partial F}{\partial a_{ij}} &= 2c_{ij} - 2\lambda A_{ij} \qquad (i, j = 1, \ldots, n;\ i < j), \end{aligned} \tag{23.8}$$

where A_{ij} $(i, j = 1, \ldots, n)$ are the cofactors of the elements a_{ij} of the matrix (a_{ij}). Equating the partial derivatives (23.8) to zero, we get

$$c_{ij} = \lambda A_{ij}, \qquad A_{ij} = \frac{c_{ij}}{\lambda} (i, j = 1, \ldots, n).$$

We consider the determinant $\det(A_{ij}/d_0)$ of the inverse of the matrix (a_{ij}). Taking advantage of the known values of the A_{ij} at the critical point on the equidiscriminant surface, we have

$$\frac{1}{d_0} = \det\left(\frac{A_{ij}}{d_0}\right) = \det\left(\frac{c_{ij}}{\lambda d_0}\right) = \frac{\det(c_{ij})}{\lambda^n d_0^n}.$$

From this relationship we find that

$$\lambda = \sqrt[n]{\frac{\det(c_{ij})}{d_0^{n-1}}} = \frac{1}{d_0} \sqrt[n]{d_0 \det(c_{ij})}.$$

We evaluate $F(a_{ij})$ at the critical point (a_{ij}^*) by substituting the numbers λA_{ij} for c_{ij} in the expression for F:

$$F(a_{ij}^*) = \sum_{i, j=1}^{n} \lambda A_{ij} a_{ij}^* - \lambda [\det(a_{ij}^*) - d_0]$$

$$= \lambda \sum_{i=1}^{n} \left(\sum_{j=1}^{n} A_{ij} a_{ij}^* \right) = \lambda n d_0 = n \sqrt[n]{d_0 \det(c_{ij})}.$$

Since the critical point is unique, it is obvious in the given case that this point is a minimum. Thus we have

$$\min M(\Gamma)=\frac{\min F(a_{ij})}{q(n+1)}=\frac{n}{q(n+1)}\sqrt[n]{d_0\det(c_{ij})}. \qquad (23.9)$$

From the relation at the minimum

$$\frac{A_{ij}}{d_0}=\frac{c_{ij}}{\lambda d_0}=\frac{c_{ij}}{\sqrt[n]{d_0\det(c_{ij})}}$$

it follows that the coordinates a_{ij}^* of this point are in fact those which were mentioned in the assertion of the proposition.

4. PROOF OF THEOREM 23.1. From (23.7) it follows that for a given lattice Γ the quantity $\max_t R_t$ in the formula (19.2) for the covering density $\theta_n(\Gamma)$ satisfies

$$\max_t R_t^2 \geqslant \max_t \frac{M_{S_t}}{n+1}.$$

Since $M(\Gamma)$ is the arithmetic mean of $M_{S_1}, \ldots, M_{S_q}$, we have

$$\max_t R_t^2 \geqslant \frac{1}{n+1} M(\Gamma)$$

and

$$\frac{\left(\max\limits_t R_t\right)^n}{\sqrt{\det(a_{ij})}} \geqslant \left[\frac{M(\Gamma)}{n+1}\right]^{n/2} \frac{1}{\sqrt{\det(a_{ij})}}.$$

Multiplying both sides of this inequality by Ω_n, taking the minimum of both sides (the minimum of either side depends only on the set of lattices of the given L-type) and making use of (23.9), we obtain the assertion of Theorem 23.1:

$$\min_L \theta_n(\Gamma) \geqslant \Omega_n \left[\frac{n}{q(n+1)^2}\right]^{n/2} \sqrt{\det(c_{ij})}.$$

5. EXAMPLE 1. We will evaluate the estimate (23.3) for Voronoĭ's type I (see [52]). For convenience we consider the case $n = 5$. In the context of arbitrary n, it is not extremely difficult to obtain the estimate (23.3) of the covering density for type I lattices, and for values $n \leqslant 4$ it is entirely trivial.

The contiguity vectors of lattices of Voronoĭ's type I are given in §3.1. They are

$$l^i,\ l^{ij},\ l^{ijk},\ l^{ijkm}\quad (i,\ j,\ k,\ m=1,\ \ldots,\ 5;\ i\neq j\neq k\neq m),$$
$$l^0=(1,\ 1,\ 1,\ 1,\ 1),$$

where the upper index denotes the numbers of the coordinates having the values 1; the remaining coordinates are equal to zero. The L-simplices of the type I lattices are given by

$$S_{ijkm}=(0,\ l^i,\ l^{ij},\ l^{ijk},\ l^{ijkm},\ l^0),\quad i,\ j,\ k,\ m=1,\ \ldots,\ 5;\ i\neq j\neq k\neq m$$

(see, for example, [2] or [10]). We will denote the squares of the lengths of these contiguity vectors by the same symbol used for the vectors themselves. The central moment of S_{ijkm}, according to (23.4), is equal to

$$M_{ijkm}=\frac{1}{6}\left(\sum_{q=1}^{5} l^{q}+l^{ij}+l^{jk}+l^{km}+l^{mp}+l^{ijk}+l^{jkm}+l^{kmp}\right.$$

$$\left.+l^{ijkm}+l^{jkmp}+l^{0}\right),\quad i,\ j,\ k,\ m,\ p=1,\ \ldots,\ 5,\quad i\neq j\neq k\neq m\neq p.$$

We obtain $M(\Gamma)$ in (23.2) as the sum of the moments M_{ijkm}:

$$M(\Gamma)=\frac{1}{360}\left(60\sum_{i=1}^{5} l^{i}+24\sum_{\substack{i,\,j=1\\(i<j)}}^{5} l^{ij}+18\sum_{\substack{i,\,j,\,k=1\\(i<j<k)}}^{5} l^{ijk}+24\sum_{\substack{i,\,j,\,k,\,m=1\\(i<j<k<m)}}^{5} l^{ijkm}+60l^{0}\right).$$

Substituting the squares of the lengths of the contiguity vectors, found by evaluating the quadratic form

$$f(x_1,\ \ldots,\ x_5)=\sum_{i,\,j=1}^{5} a_{ij}x_ix_j,$$

into the expression for $M(\Gamma)$, we have

$$M(\Gamma)=\frac{1}{360}\left(420\sum_{i=1}^{5} a_{ii}+210\sum_{\substack{i,\,j=1\\(i\neq j)}}^{5} a_{ij}\right),$$

from which we obtain

$$c_{ii}=420\ (i=1,\ldots,\ 5),\quad c_{ij}=210\ (i,\ j=1,\ldots,\ 5;\ i<j).$$

According to (23.3), we have

$$\det(c_{ij})=210^5\cdot 6,\qquad M_5=\Omega_5\left(\frac{35}{72}\right)^{5/2}\sqrt{6}.$$

By comparing M_5 with the covering density θ_5^0 of the lattice Γ_5 (§19, Theorem T_n) we see that they are equal. Having in mind that our results are correct not only for $n = 5$ but for all n, we deduce from Theroem 23.1 that the minimum covering densities for lattices of Voronoĭ's type I are achieved at the lattices Γ_n and are equal to θ_n^0.

As one would expect, and as is easy to prove using Proposition 23.3, the minimum of $M(\Gamma)$, yielding the bound M_n, is achieved on the lattice Γ_n and is equal to θ_n^0. This coincidence can be explained by the fact that the centers of mass of the L-simplices of Γ_n coincide with the centers of their circumscribed spheres.

EXAMPLE 2. *The bound* (23.3) *for types* II *and* III *when* $n = 4$. Knowing the L-simplices of the four-dimensional lattices of types II and III (see for example [2], [10], [22], or §9 of this work), we can find expressions for the squares of the lengths of their edges using the coefficients of their respective quadratic forms. Using (23.2), we are able to calculate $M(\Gamma)$ for each of these types:

$$M_{II}(\Gamma)=\frac{1}{60}\left[30\left(\sum_{i,j=1}^{4}a_{ij}+\sum_{i=1}^{4}a_{ii}\right)-36a_{12}\right],$$
$$M_{III}(\Gamma)=\frac{1}{60}\left[30\left(\sum_{i,j=1}^{4}a_{ij}+\sum_{i=1}^{4}a_{ii}\right)-36a_{12}-24a_{34}\right]. \tag{23.10}$$

We remark that in [22] the expression (23.10) is presented in a useful form. In order to use the results of §9 of the present work it is necessary to neglect the transformation from the parameters ρ_{ij} and $\rho_{\overline{ij}}$ to the coefficients a_{ij} of the quadratic form.

The expressions (23.10) for $M_{II}(\Gamma)$ and $M_{III}(\Gamma)$ yield the following values for the coefficients c_{ij} for each of the types under consideration:

II. $c_{ii}=60,\ c_{ij}=30\ (i,j=1,\dots,4;\ i\neq j;\ ij\neq 12),\ c_{12}=12;$
III. $c_{ii}=60,\ c_{ij}=30\ (i,j=1,\dots,4;\ i\neq j;\ ij\neq 12,34),\ c_{12}=12,\ c_{34}=18.$

Using the coefficients c_{ij} we find that

$$\det{}_{II}(c_{ij})=8\cdot 720^2,\qquad \det{}_{III}(c_{ij})=12^4\cdot 14^2,$$

and we can calculate the bounds (23.3):

$$M_{II}=\Omega_4\frac{32\sqrt{2}}{125}=\Omega_4\frac{4}{25}\sqrt{5.12},\qquad M_{III}=\Omega_4\frac{224}{625}=\Omega_4\frac{4}{25}\sqrt{5.0176}.$$

Comparing these estimates with the value $\theta_4^0=\Omega_4\cdot 4\sqrt{5}/25$ and relying on the result of Example 1 for type I, we have a proof of Theorem T_4.

EXAMPLE 3. We will consider the L-type described in [22] and referred to as type IV (for $n=5$ it is the key type 60). Type IV occurs for all $n\geqslant 5$; a description of the L-subdivision and the calculation of $M(\Gamma)$ is given in [22] for all these values of n. (In [22] there are some misprints: in the numerator of the formula for Y_I it is necessary to replace $(n+1)!$ by $(n+2)!$, in the denominator of the formula for Y_{III} replace $(n+2)$ by $(n-2)$, and in the formula for $(\min Y_{III}/Y_I)^n$ replace the coefficient "–2998" by "–2298"). Using [22], for the coefficients c_{ij} we obtain

$$c_{ii}=\frac{(n+2)!}{12},\qquad c_{ij}=\frac{(n+2)!}{24}\qquad (i,j=1,\dots,n;\ i\neq j;\ ij\neq 12,\ 23),$$
$$c_{12}=\frac{(n+2)!}{24}\left[1-\frac{12}{n(n+1)}\right],\qquad c_{23}=\frac{(n+2)!}{24}\left[1-\frac{24}{n(n^2-1)}\right].$$

After some calculation we have

$$\det(c_{ij})=\left[\frac{(n+2)!}{24}\right]^n(n+1+\omega),\qquad M_n=\Omega_n\left[\frac{n(n+2)}{12(n+1)}\right]^{n/2}\sqrt{n+1+\omega},$$

where

$$\omega=\frac{24(n-3)(n^3-7n-18)}{(n^2-1)^2n^2}.$$

Comparing the value M_n with the value θ_n^0 ($n \geqslant 5$), we see that $M_n > \theta_n^0$, from which it follows that lattices of type IV give greater densities than do the lattices Γ_n. This statement is proved in [22] for $n \geqslant 6$; for $n = 5$ it is proved in [23] by means of a direct calculation of the minimum density for the set of type IV lattices.

§24. Proof of Lemma 3

1. We can prove Lemma 3 if we show that the lower bound M_5 of Theorem 23.1 for the covering density for each of the key L-types 8–81 of the domain R_1 is greater than the density θ_5^0. We will show that this is so.

In order to find, for a given L-type, the estimate (23.3) of the covering density M_n, we must know the elements defining $\det(c_{ij})$, i.e. we need to find the central moment $M(\Gamma)$ (see (23.2)) for lattices of this L-type.

We turn to the L-types of R_1. All the investigations of the L-types of this domain were carried out using the parameters ρ_{ij} and $\rho_{\overline{ij}}$ appearing in (3.4). In particular, the squares of the lengths of the contiguity vectors were calculated using these parameters (15.7). For the calculation of the coefficients c_{ij} in the expression (23.2) for $M(\Gamma)$ we require the squares of the lengths of the contiguity vectors expressed in terms of the coefficients a_{ij} of the quadratic form.

We handle this situation in the following way: all of the calculations involving $M(\Gamma)$ for lattices of a given L-type will be carried out with the parameters ρ_{ij} and $\rho_{\overline{ij}}$. Thus we will obtain (for $n = 5$ and $q = 60$) a formula

$$M(\Gamma) = \frac{1}{360}\left[\sum_{\substack{i,j=1\\(i<j)}}^{5} (\alpha_{ij}\rho_{ij} + \alpha_{\overline{ij}}\rho_{\overline{ij}})\right], \tag{24.1}$$

which is the analogue of (23.2). Then we will find the values of the coefficients c_{ij} of the right side (23.2) expressed in terms of α_{ij} and $\alpha_{\overline{ij}}$ as a preliminary. We pass now to the derivation of these expressions.

From (3.4) we find that

$$\begin{aligned}
&a_{11} = \rho_{12} + \rho_{13} + \rho_{14} + \rho_{15} + \rho_{23} + \rho_{24} + \rho_{25} + \rho_{34} + \rho_{35} + \rho_{45},\\
&a_{22} = \rho_{\overline{12}} + \rho_{\overline{23}} + \rho_{\overline{24}} + \rho_{\overline{25}} + \rho_{13} + \rho_{14} + \rho_{15} + \rho_{34} + \rho_{35} + \rho_{45},\\
&a_{rr} = \rho_{1r} + \rho_{2r} + \rho_{rs} + \rho_{rt} + \rho_{\overline{1r}} + \rho_{\overline{2r}} + \rho_{\overline{rs}} + \rho_{\overline{rt}},\\
&a_{12} = \rho_{13} + \rho_{14} + \rho_{15} + \rho_{34} + \rho_{35} + \rho_{45},\\
&a_{1r} = -\rho_{1r} - \rho_{2r} - \rho_{rs} - \rho_{rt}, \qquad a_{2r} = -\rho_{1r} - \rho_{\overline{2r}} - \rho_{rs} - \rho_{rt},\\
&a_{rs} = -\rho_{\overline{rs}} + \rho_{rs} \qquad (r,\ s,\ t = 3,\ 4,\ 5;\ r \neq s \neq t).
\end{aligned}$$

Substituting these values a_{ij} into the right-hand side of (23.2) and comparing it with the right-hand side of (24.1), we find expressions for α_{ij} and $\alpha_{\overline{ij}}$ in terms of c_{ij}, and from these we calculate the c_{ij}:

$$c_{11}=\alpha_{12},\qquad c_{mm}=\alpha_{\overline{1m}}\qquad (m=2,\ 3,\ 4,\ 5),$$

$$c_{1r}=\frac{1}{2}(\alpha_{12}+\alpha_{\overline{1r}}-\alpha_{2r}),\qquad c_{2r}=\frac{1}{2}(\alpha_{\overline{12}}+\alpha_{\overline{1r}}-\alpha_{\overline{2r}})\qquad (r=3,\ 4,\ 5),$$

$$c_{34}=\frac{1}{2}(\alpha_{\overline{13}}+\alpha_{\overline{14}}-\alpha_{\overline{34}}),\qquad c_{35}=\frac{1}{2}(\alpha_{\overline{13}}+\alpha_{\overline{15}}-\alpha_{\overline{35}}),\qquad c_{45}=\frac{1}{2}(\alpha_{\overline{14}}+\alpha_{\overline{15}}-\alpha_{45}),$$

$$c_{12}=\frac{1}{2}(\alpha_{13}+\alpha_{\overline{13}}-\alpha_{23}-\alpha_{\overline{23}})=\frac{1}{2}(\alpha_{14}+\alpha_{\overline{14}}-\alpha_{42}-\alpha_{\overline{24}})=\dots. \tag{2.42}$$

The possibility of calculating c_{12} using several (six) different formulas can be explained by the nonuniqueness of the parameters ρ_{ij} and $\rho_{\overline{ij}}$ used to express the quadratic form.

2. We proceed to calculate the quantities M_5 for each key L-type of the domain R_1. The first step in the calculation of M_5 is the determination of the coefficients α_{ij} and $\alpha_{\overline{ij}}$ in (24.1).

As a preliminary we calculate the mean of the central moments for a block of L-simplices of a key L-type. Making use of (23.6) and (15.7), we enumerate the simplices of a block and the edges of these simplices (Figures 16–18). For the block $\{ij\}$ we obtain

$$M(\{ij\})=\frac{1}{36}\Big[30(\rho_{ij}+\rho_{\overline{ij}})+92(\rho_{km}+\rho_{kp}+\rho_{mp})+36(\rho_{\overline{km}}+\rho_{\overline{kp}}+\rho_{\overline{mp}})$$
$$+47\sum_{\mu=k,\,m,\,p}(\rho_{i\mu}+\rho_{j\mu}+\rho_{\overline{i\mu}}+\rho_{\overline{j\mu}})\Big], \tag{24.3}$$

for the block $[ij,\ k]$

$$M([ij,\ k])=\frac{1}{36}\Big[30(\rho_{ij}+\rho_{\overline{ij}})+64(\rho_{mp}+\rho_{\overline{mp}})+92(\rho_{kp}+\rho_{km})$$
$$+36(\rho_{\overline{kp}}+\rho_{\overline{km}})+47\sum_{\mu=k,\,m,\,p}(\rho_{i\mu}+\rho_{j\mu}+\rho_{\overline{i\mu}}+\rho_{\overline{j\mu}})\Big], \tag{24.4}$$

and for the block $(ij,\ km)$

$$M(ij,\ km)=\frac{1}{12}\Big[24(\rho_{ij}+\rho_{km})+10(\rho_{\overline{ij}}+\rho_{\overline{km}})+13\sum_{\mu=i,\,j,\,k,\,m}(\rho_{\mu p}+\rho_{\overline{\mu p}})$$
$$+17\sum_{\substack{\mu=i,\,j;\\ \sigma=k,\,m}}(\rho_{\mu\sigma}+\rho_{\overline{\mu\sigma}})\Big]. \tag{24.5}$$

After the coefficients α_{ij} and $\alpha_{\overline{ij}}$ are found, we must use (24.2) and then (23.3), which in the case under consideration ($n = 5, q = 60$) has the form

$$M_5=\Omega_5\left(\frac{1}{12\cdot 6^2}\right)^{5/2}\sqrt{\det(c_{ij})}.$$

For convenience we will find the quantities

$$\mu_\nu=\left(\frac{M_5}{\theta_5^0}\right)^2=\frac{\det(c_{ij})}{6\,(210)^5} \tag{24.6}$$

for the key L-types contained in the C-types with diagrams $\mathfrak{S}_\nu$ ($\nu = 1, \dots, 81$).

Naturally we will not reproduce here the calculations for each of the values μ_ν, but will restrict ourselves to the consideration of some examples. All of the quantities

μ_ν, $\nu = 1, \ldots, 81$, are given in Table VII.

TABLE VII

The values of μ_ν for the key L-types of R_1

ν	μ_ν	ν	μ_ν	ν	μ_ν	ν	μ_ν
1	1,00000000	22	1,01376000	42	1,03196444	62	1,05801482
2	0,99200000	23	1,04521480	43	1,04530962	63	1,05411686
3	0,98933333	24	1,03632591	44	1,03480888	64	1,07034074
4	0,98773333	25	1,04640000	45	1,04000000	65	1,05327407
5	0,96762360	26	1,04106666	46	1,03359999	66	1,07223704
6	0,97920000	27	1,03881481	47	1,03205926	67	1,08266666
7	0,98304000	28	1,03006814	48	1,06370370	68	1,08835556
8	1,02666666	29	1,06666666	49	1,06939259	69	1,07887407
9	1,01653333	30	1,06986666	50	1,05837037	70	1,07971687
10	1,00533333	31	1,06477037	51	1,05991112	71	1,05422222
11	1,00608000	32	1,07152593	52	1,04107851	72	1,05590781
12	1,03472592	33	1,05801481	53	1,04816894	73	1,02666666
13	1,03170370	34	1,04770370	54	1,04964214	74	1,03703703
14	1,02761481	35	1,04637900	55	1,04964214	75	1,03242799
15	1,02821926	36	1,05664422	56	1,05882073	76	1,03620741
16	1,02189037	37	1,04106666	57	1,04761152	77	1,02376297
17	1,07361717	38	1,04469333	58	1,06316116	78	1,02059193
18	1,00770370	39	1,04599704	59	1,05197037	79	1,03316543
19	1,04000000	40	1,05303704	60	1,04000000	80	1,03552526
20	1,03253333	41	1,04241777	61	1,04853333	81	1,04127737
21	1,02080000						

EXAMPLE 1. We will find μ_1 for the Barnes type (L-type 1 of R_1). The blocks of the L-subdivision of lattices of this type are given in Example 1 of §15. Making use of (24.5), we obtain

$$M(\Gamma)=\frac{1}{360}84\left[6(\rho_{12}+\ldots+\rho_{45})+5(\rho_{\overline{12}}+\ldots+\rho_{\overline{45}})\right].$$

Further we find the quantities c_{ij} by (24.2) and calculate the determinant

$$\det(c_{ij})=(42)^5\begin{vmatrix}12&0&5&5&5\\0&10&5&5&5\\5&5&10&5&5\\5&5&5&10&5\\5&5&5&5&10\end{vmatrix}=(210)^5\cdot 6.$$

From (24.6) we find that $\mu_1 = 1$.

EXAMPLE 2. *Key L-type* 5. The blocks for this L-subdivision are given in Example 2 of §15. By means of the transformation

$$x_i\to -x_i,\ x_j\to x_j,\ x_k\to -x_k,\ x_m\to x_m,\ x_p\to x_p$$

we modify (24.5), obtaining

$$M(\overline{ij},\ \overline{km})=\frac{1}{12}\Big[24(\rho_{\overline{ij}}+\rho_{\overline{km}})+10(\rho_{ij}+\rho_{km})$$
$$+13\sum_{\mu=i,\,j,\,k,\,m}(\rho_{\mu p}+\rho_{\overline{\mu p}})+17\sum_{\mu=i,\,j;\ \sigma=k,\,m}(\rho_{\mu\sigma}+\rho_{\overline{\mu\sigma}})\Big],$$

which we use to obtain

$$M(\Gamma)=\frac{1}{360}84\,[5(\rho_{12}+\dots+\rho_{45})+6(\rho_{\overline{12}}+\dots+\rho_{\overline{45}})].$$

Using (24.2) and the values $\alpha_{ij}=5\cdot 84$ and $\alpha_{\overline{ij}}=6\cdot 84$ $(i,j=1,\dots,5;\,i<j)$ we find the coefficients c_{ij} and then evaluate the determinant:

$$\det(c_{ij})=42^5\begin{vmatrix}10&0&6&6&6\\0&12&6&6&6\\6&6&12&6&6\\6&6&6&12&6\\6&6&6&6&12\end{vmatrix}=42^5\cdot 6^4\cdot 14.$$

Formula (24.6) yields

$$\mu_5=\frac{36^2\cdot 7}{25^2\cdot 15}=\frac{9072}{9375}<1.$$

EXAMPLE 3. *The key L-type* 8. From one of its blocks $\{12\}$ and from (24.3) we find that

$$M(\{12\})=\frac{1}{36}\left[30(\rho_{12}+\rho_{\overline{12}})+\sum_{\substack{i,j=3\\(i<j)}}^{5}(92\rho_{ij}+36\rho_{\overline{ij}})+47\sum_{i=3}^{5}(\rho_{1i}+\rho_{\overline{1i}}+\rho_{2i}+\rho_{\overline{2i}})\right].$$

The central moments of the remaining blocks are calculated by means of (24.5). As a result we have

$$M(\Gamma)=\frac{1}{360}\left[5\rho_{12}+5(\rho_{\overline{12}}+\dots+\rho_{\overline{45}})+6\sum_{i=3}^{5}(\rho_{1i}+\rho_{2i})+7(\rho_{34}+\rho_{35}+\rho_{45})\right],$$

$$\det(c_{ij})=42^5\begin{vmatrix}10&0&4&4&4\\0&10&5&5&5\\4&5&10&5&5\\4&5&5&10&5\\4&5&5&5&10\end{vmatrix}=42^5\cdot 5^3\cdot 154$$

and $\mu_8=77/75>1$.

3. In order to establish the validity of Lemma 3 it remains to consider the results in Table VII. From Theorem 23.1 and from (24.6) it follows that if for the key type ν the quantity μ_ν has a value greater than 1, then the covering densities for all lattices of this type are greater than θ_5^0. Since $\mu_\nu\geqslant 1$ in Table VII for all $\nu\geqslant 8$, this finishes the proof of Lemma 3 and the proof of Theorem T_5.

BIBLIOGRAPHY

1. G. F. Voronoĭ, *Nouvelles applications des paramètres continus à la théorie des formes quadratiques. Premier mémoire*, J. Reine Angew. Math. **133** (1908), 79–178.

2. ———, *Nouvelles applications des paramètres continus à la théorie des formes quadratiques. Deuxieme mémoire*, J. Reine Angew. Math. **134** (1908), 198–287; **136** (1909), 67–178.

3. S. S. Ryškov, *The C-type of n-dimensional parallelehedra*, Dokl. Akad. Nauk SSSR **212** (1973), 46–49; English transl. in Soviet Math. Dokl. **14** (1973).

4. E. P. Baranovskiĭ and S. S. Ryškov, *Primitive five-dimensional parallelehedra*, Dokl. Akad. Nauk SSSR **212** (1973), 532–535; English transl. in Soviet Math. Dokl. **14** (1973).

5. ———, *The combinatorial-metric structure of L-partitions of general five-dimensional lattices*, Dokl. Akad. Nauk SSSR **220** (1975), 265–268; English transl. in Soviet Math. Dokl. **16** (1975).

6. S. S. Ryškov and E. P. Baranovskiĭ, *Solution of the problem of least dense lattice covering of five-dimensional space by equal spheres*, Dokl. Akad. Nauk SSSR **222** (1975), 39–42; English transl. in Soviet Math. Dokl. **16** (1975).

7. E. S. Fedorov, *Elements of the study of figures*, Zap. Mineralog. Obšč. (2) **21** (1885), 1–279; reprint, Izdat. Akad. Nauk SSSR, Moscow, 1953. (Russian)

8. A. D. Aleksandrov, *Déduction des paralléloèdres non normaux à 4 dimensions*, Izv. Akad. Nauk SSSR Otdel. Mat. Estestven. Nauk (7) **1934**, 803–817. (Russian; French summary)

9. G. Lejeune Dirichlet, *Über die Reduction der positiven quadratischen Formen mit drei unbestimmten ganzen Zahlen*, J. Reine Angew. Math. **40** (1850), 209–227; reprinted in his *Werke*, Band II, Georg Reimer, Berlin, 1897; Chelsea, New York, 1964, pp. 27–48.

10. B. N. Delone, *Geometry of positive quadratic forms*, Uspehi Mat. Nauk **3** (1937), 16–62; **4** (1938), 102–164. (Russian)

11. ———, *Sur la partition régulière de l'espace à 4 dimensions*, Izv. Akad. Nauk SSSR Otdel. Fiz.-Mat. Nauk (7) **1929**, 79–110, 147–164.

12. O. K. Žitomirskiĭ, *Verschärfung eines Satzes von Voronoi*, Ž. Leningrad. Fiz.-Mat. Obšč. **2** (1929), no. 2, 131–151 (1931).

13. B. Delaunay [Delone], *Sur la sphère vide*, Proc. Internat. Congr. Math. (Toronto, 1924), vol. 1, Univ. of Toronto Press, Toronto, 1928, pp. 695–700.

14. ———, *Sur la sphère vide*, Izv. Akad. Nauk SSSR Otdel. Mat. Estestven. Nauk (7) **1934**, 793–800.

15. B. N. Delone and N. N. Sandakova, *Theory of stereohedra*, Trudy Mat. Inst. Steklov. **64** (1961), 28–51. (Russian)

16. Hermann Minkowski, *Allgemeine Lehrsätze über die konvexen Polyeder*, Nachr. Ges. Wiss. Göttingen Math.-Phys. Kl. **1897**, 198–219; reprinted in his *Gesammelte Abhandlungen*, Band II, Teubner, Leipzig, 1911, pp. 103–121.

17. ———, *Diophantische Approximationen. Eine Einführung in die Zahlentheorie*, Teubner, Leipzig, 1907; reprint, Chelsea, New York, 1957.

18. B. A. Venkov, *On the projection of parallelehedra*, Mat. Sb. **49 (91)** (1959), 207–224. (Russian)

19. B. N. Delone and S. S. Ryškov, *Solution of the problem of least dense lattice covering of a four-dimensional space by equal spheres*, Dokl. Akad. Nauk SSSR **152** (1963), 523–524; English transl. in Soviet Math. Dokl. **4** (1963).

20. E. S. Barnes and T. J. Dickson, *Extreme coverings of n-space by spheres*, J. Austral. Math. Soc. **7** (1967), 115–127.

21. B. N. Delone et al., *A new construction of the theory of lattice coverings of an n-dimensional space by equal spheres*, Izv. Akad. Nauk SSSR Ser. Mat. **34** (1970), 289–298; English transl. in Math. USSR Izv. **4** (1970).

22. S. S. Ryškov, *Some remarks on the types of n-dimensional parallelehedra and on the density of lattice coverings of the space E^n by equal spheres*, Dokl. Akad. Nauk SSSR **162** (1965), 277–280; English transl. in Soviet Math. Dokl. **6** (1965).

23. D. W. Trenerry, *The covering of space by spheres*, Doctoral Dissertation, University of Adelaide, Adelaide, 1972.

24. T. J. Dickson, *On Voronoi reduction of positive definite quadratic forms*, J. Number Theory **4** (1972), 330–341.

25. Ch. Hermite, *Extraits de lettres de M. Ch. Hermite à M. Jacobi sur différents objets de la théorie des nombres*, J. Reine Angew. Math. **40** (1850), 261–315; reprinted in his *Oeuvres*, Tome I, Gauthier-Villars, Paris, 1905, pp. 100–163.

26. B. A. Venkov, Commentary on [1], in G. F. Voronoĭ, *Collected works*. Vol. 2, Izdat. Akad. Nauk Ukrain. SSR, Kiev, 1952, pp. 379–385. (Russian)

27. ———, *Über die Reduktion positiver quadratischer Formen*, Izv. Akad. Nauk SSSR Ser. Mat. **4** (1940), 37–52. (Russian; German summary)

28. S. S. Ryškov, *The Venkov reduction of positive quadratic forms*, Ivanovo Gos. Univ. Učen. Zap. **89** (1974), 5–36. (Russian)

29. Hermann Minkowski, *Diskontinuitätsbereich für arithmetische Äquivalenz*, J. Reine Angew. Math. **129** (1905), 220–294; reprinted in *Gesammelte Abhandlungen*, Band 2, Teubner, Leipzig, 1911, pp. 53–100.

30. B. L. van der Waerden, *Die Reduktionstheorie der positiven quadratischen Formen*, Acta Math. **96** (1956), 265–309.

31. S. S. Ryškov, *On the theory of the reduction of positive quadratic forms in the sense of Hermite and Minkowski*, Zap. Naučn. Sem. Leningrad. Otdel. Mat. Inst. Steklov. (LOMI) **33** (1973), 37–64; English transl. in J. Soviet Math. **6** (1976).

32. N. N. Sandakova, *On Dirichlet partitions for n-dimensional crystallographic space groups*, Candidate's Dissertation, Steklov Inst. Math. Acad. Sci. USSR, Moscow, 1959. (Russian)

33. B. N. Delone, S. S. Ryškov and M. I. Štogrin, *On a theorem of Sandakova in the theory of positive quadratic forms*, Mat. Zametki **1** (1967), 253–262; English transl. in Math. Notes **1** (1967).

34. E. P. Baranovskiĭ, *A theorem on L-partitions of lattices of points*, Mat. Zametki **13** (1973), 605–616; English transl. in Math. Notes **13** (1973).

35. S. S. Ryškov, *The basic extremal problems of the geometry of positive quadratic forms*, Doctoral Dissertation, Steklov Inst. Math. Acad. Sci. USSR, Moscow, 1970. (Russian)

36. ———, *The perfect form A_n^k*, Zap. Naučn. Sem. Leningrad. Otdel. Mat. Inst. Steklov. (LOMI) **33** (1973), 65–71; English transl. in J. Soviet Math. **6** (1976).

37. E. P. Baranovskiĭ, *Volumes of L-simplices of five-dimensional lattices*, Mat. Zametki **13** (1973), 771–782; English transl. in Math. Notes **13** (1973).

38. S. S. Ryškov, *The structure of an n-dimensional parallelehedron of the first kind*, Dokl. Akad. Nauk SSSR **146** (1962), 1027–1030; English transl. in Soviet Math. Dokl. 3 (1962).

39. E. S. Barnes and D. W. Trenerry, *A class of extreme lattice coverings of n-space by spheres*, J. Austral. Math. Soc. **14** (1972), 247–256.

40. E. P. Baranovskiĭ, *Simplices of L-decompositions of Euclidean spaces*, Mat. Zametki **10** (1971), 659–670; English transl. in Math. Notes **10** (1971).

41. S. S. Ryškov, *Effectuation of a method of Davenport in the theory of coverings*, Dokl. Akad. Nauk SSSR **175** (1967), 303–305; English transl. in Soviet Math. Dokl. 8 (1967).

42. Richard Kershner, *The number of circles covering a set*, Amer. J. Math. **61** (1939), 665–671.

43. R. P. Bambah, *On lattice coverings by spheres*, Proc. Nat. Inst. Sci. India **20** (1954), 25–52.

44. E. S. Barnes, *The covering of space by spheres*, Canad. J. Math. **8** (1956), 293–304.

45. E. P. Baranovskiĭ, *Local density minima of a lattice covering of a four-dimensional Euclidean space by equal spheres*, Dokl. Akad. Nauk SSSR **164** (1965), 13–15; English transl. in Soviet Math. Dokl. **6** (1965).

46. E. P. Baranovskiĭ and S. S. Ryškov, *The second local density minimum of a lattice covering of a four-dimensional Euclidean space by equal spheres*, Sibirsk. Mat. Ž. **7** (1966), 731–739; English transl. in Siberian Math. J. (1966).

47. E. P. Baranovskiĭ, *Local minima of the density of a lattice covering of a four-dimensional Euclidean space by equal spheres*, Sibirsk. Mat. Ž. **7** (1966), 974–1001; English transl. in Siberian Math. J. **7** (1966).

48. T. J. Dickson, *An extreme covering of 4-space by spheres*, J. Austral. Math. Soc. **6** (1966), 179–192.

49. ———, *The extreme coverings of 4-space by spheres*, J. Austral. Math. Soc. **7** (1967), 490–496.

50. M. N. Bleicher, *Lattice coverings of n-space by spheres*, Canad. J. Math. **14** (1962), 632–650.

51. A. F. Gameckiĭ, *On the theory of covering an n-dimensional Euclidean space by equal spheres*, Dokl. Akad. Nauk SSSR 146 (1962), 991–994; English transl. in Soviet Math. Dokl. **3** (1962).

52. ———, *Optimality of the Voronoĭ principal lattice of first type among the lattices of first type of any number of dimensions*, Dokl. Akad. Nauk SSSR **151** (1963), 482–484; English transl. in Soviet Math. Dokl. **4** (1963).

53. B. N. Delone and S. S. Ryškov, *Extremal problems in the theory of positive quadratic forms*, Trudy Mat. Inst. Steklov. **112** (1971), 203–223; English transl. in Proc. Steklov Inst. Math. **112** (1971).

54. G. M. Fihtengol'c, *A course in differential and integral calculus*. Vol. I, Fizmatgiz, Moscow, 1958; German transl., VEB Deutscher Verlag, Berlin, 1968.

55.* H. S. M. Coxeter, *The classification of zonohedra by means of projective diagrams*, J. Math. Pures Appl. (9) **41** (1962), 137–156.

---

* *Added in translation.*